Helena Banning

Selenium Isotope Studies in Plants

Development and Validation of a Novel Geochemical Tool and its Application to Organic Samples

AF875972

Karlsruher Mineralogische und Geochemische Hefte
Schriftenreihe des Instituts für Angewandte Geowissenschaften,
Karlsruher Institut für Technologie (KIT)

Band 42

Selenium Isotope Studies in Plants

Development and Validation of a Novel Geochemical Tool and its Application to Organic Samples

by
Helena Banning

KIT Scientific Publishing

Dissertation, Karlsruher Institut für Technologie (KIT)
Fakultät für Bauingenieur-, Geo- und Umweltwissenschaften, 2016
Referenten: Prof. Dr. Thomas Neumann, Prof. Dr. Ronny Schönberg

Impressum

KIT Scientific Publishing

Karlsruher Institut für Technologie (KIT)
KIT Scientific Publishing
Straße am Forum 2
D-76131 Karlsruhe

KIT Scientific Publishing is a registered trademark of Karlsruhe Institute of Technology. Reprint using the book cover is not allowed.

www.ksp.kit.edu

This document – excluding the cover, pictures and graphs – is licensed under the Creative Commons Attribution-Share Alike 3.0 DE License (CC BY-SA 3.0 DE): http://creativecommons.org/licenses/by-sa/3.0/de/

The cover page is licensed under the Creative Commons Attribution-No Derivatives 3.0 DE License (CC BY-ND 3.0 DE): http://creativecommons.org/licenses/by-nd/3.0/de/

Print on Demand 2017 – Gedruckt auf FSC-zertifiziertem Papier

ISSN 1618-2677
ISBN 978-3-7315-0594-5
DOI: 10.5445/KSP/1000060292

Selenium Isotope Studies in Plants

Development and Validation of a Novel Geochemical Tool and its Application to Organic Samples

submitted in fulfilment of the requirements for the degree of

Doctor of Natural Sciences (Dr. rer. nat.)

To the Department of Civil Engineering, Geo- and Environmental Sciences

approved

Dissertation

of

Helena Banning

born in Steinfurt

Day of oral examination: 17. February 2016

Reviewer: Prof. Dr. Thomas Neumann

Second reviewer: Prof. Dr. Ronny Schönberg

Advisor: Dr. Monika Stelling

Kurzfassung

Selen (Se) ist ein essentieller Nährstoff und toxischer Schadstoff mit einer engen Toleranzgrenze. Aufgrund von heterogener Verteilung auf lokaler und globaler Ebene beinhaltet eine inadäquate Selenversorgung ein hohes Risiko, welches 0,5 bis eine Milliarde Menschen weltweit betrifft. Pflanzen sind für den Menschen die Hauptquelle für Selen und gleichzeitig der praktisch einzige Weg zum Eintritt des Selens in die Nahrungskette. Redoxprozesse bestimmen maßgeblich die Aufnahme und Verteilung von Selen innerhalb der Pflanze. Selenisotopenvariationen in geologischen Systemen und der oberflächennahen Umwelt wurden als zuverlässige Redoxtracer erkannt, was deren Anwendung für die Erforschung des Selenkreislaufs in Pflanzen äußerst vielversprechend macht. Dennoch sind Daten zur Selenisotopenverteilung in Pflanzen knapp und die Komplexität natürlicher Systeme macht eine differenzierte Untersuchung einzelner Prozesse kaum möglich. Ziel dieser Studie ist die getrennte Untersuchung von selenbezogenen Prozessen in Pflanzen und die Aufdeckung der Zusammenhänge zwischen diesen Prozessen und Selenisotopenvariationen, die durch jene ausgelöst wurden. Zu diesem Zweck wurde ein *Minimum Parameter* Ansatz gewählt, in dem Pflanzen in geschlossenen und kontrollierten Systemen unter Ausschluss externer Einflüsse gezüchtet wurden. Phytoagar, ein künstliches nährstofffreies Wachstumsmedium, wurde mit Selen in verschiedenen Spezies und Konzentrationen dotiert, und Pflanzen wurden innerhalb einer geschlossenen Box darin kultiviert. Massenbilanzierung und die Bestimmung von Selenisotopenverhältnissen in den Kompartimenten ermöglichten die Quantifizierung von Aufnahme, Translokation und Volatilisierung sowie Isotopenfraktionierungen, die im Rahmen dieser Prozesse ausgelöst wurden. Selenisotopenverhältnisse (δ^{82}Se) wurden mittels Hydridgeneration-Multikollektor-Induktiv gekoppeltem Plasma-Massenspektrometrie (HG-MC-ICP-MS) bestimmt ($\delta^{82}Se\ [‰] = ((^{82/76}Se_{Probe})/(^{82/76}Se_{Standard})-1)*1000$). Diese Technik zeichnet sich durch hohe Präzision und gleichzeitig hohe Empfindlichkeit auf Matrixeffekte aus. Akkurate und valide Daten erfordern daher mehrere Probenaufbereitungsschritte. Aus dem Grund wurde in dieser Arbeit eine umfassende Methode für die sinnvolle, präzise und valide Bestimmung von stabilen Selenisotopen in Pflanzen und Phytoagar entwickelt. Grundlage dafür war die Transformation beider Probenarten in eine flüssige, möglichst organikfreie Form. Darauf aufbauend wurden die Komponenten der Probenmatrix selektiv vom Proben-Selen getrennt, um Masseninterferenzen und matrixbedingte Störungen während der Analysen zu vermeiden. Für die jeweiligen Schritte und Probenarten wurden diverse Methoden angepasst oder neu entwickelt und systematisch auf Effektivität, Validität und mögliche Einschränkungen getestet.

Eine mikrowellenbasierte Aufschlussmethode mit zwei getrennten Reaktionskammern wurde als das am besten geeignete Verfahren für Pflanzenmaterial bewertet, da sie eine Mineralisierungsrate von 99.3 (±0.4) % und eine reproduzierbar komplette Wiedergewinnung des Selens garantiert. Für die Behandlung des Phytoagars wurde eine speziell entwickelte Vakuumfiltrationsmethode für am geeignetsten erachtet, die jedoch einige Limitierungen bezüglich organischer Reste aufwies. Die Anwendung von Anionenaustausch im Säulentrennverfahren zur Entfernung der Probenmatrix zeigte eine hohe Matrixabhängigkeit, die für selektive Retention durch Thiolgruppen nicht entdeckt werden konnte. Beide Methoden ermöglichten die vollständige Entfernung kritischer Matrixelemente, jedoch blieben in beiden Fällen organische Phasen zusammen mit Selen zurück. Selenabtrennung durch Hydridbildung hingegen konnte Selen vollständig von der organischen Matrix abtrennen. Validierungstests zeigten, dass organische Reste einen schwerwiegenden Einfluss auf die Korrektur instrumenteller Fraktionierung und damit auf die Validität der Ergebnisse haben. Daher erfüllten Anionenaustausch und Thiolretention nicht die analytischen Voraussetzungen. Im Gegensatz dazu brachte die Hydridseparation valide und verlässliche Ergebnisse mit einer δ^{82}Se-Präzision von 0.2 (±0.2) ‰ für Pflanzen und 1.1 (±0.1) ‰ für Phytoagar hervor. Die Validität dieser Methode wurde über die Messung von zertifiziertem organikreichem Tonstein (SGR-1) bestätigt. Die hier entwickelte und validierte Methode bietet eine solide und verlässliche Basis für die Messung stabiler Selenisotope in organischen Proben und charakterisiert organische Komponenten als limitierenden Faktor für valide Analytik. Auf dieser Grundlage wurden Selenisotopenvariationen innerhalb der Kompartimente des *Minimum Parameter* Ansatzes detektiert. Transportprozesse offenbarten eine große Abhängigkeit von Aufnahme und Translokation gegenüber der Selenquellspezies sowie charakteristische limitierende Schritte, die besonders bei hoher Selenexposition auftraten. Trotz hoher Umsatzraten führten weder Aufnahme noch Volatilisierung zu signifikanter Selenisotopenfraktionierung. Translokation hingegen induzierte bedeutende Fraktionierungen ($\Delta^{82}Se$) von +2.3 ‰ bis +3.5 ‰ in Selenat beziehungsweise +1.2 ‰ bis +1.9 ‰ in Selenit dotierten Ansätzen. Höhere Konzentrationen im Wachstumsmedium gingen dabei mit geringeren Fraktionierungswerten einher. Diese Resultate indizieren eine Verschiebung von metabolischen Selenumsatzmechanismen, die Schlüsselprozesse für die Selenakkumulation in bestimmten Pflanzenteilen darstellen. Kenntnisse darüber sind von essentieller Relevanz zur Erforschung und Sicherstellung adäquater Selenversorgung durch pflanzliche Ernährung.

Diese Arbeit bietet eine umfassende und validierte Methode zur Selenisotopenanalytik in organischen Proben mit einer Präzision, die zur Detektion pflanzeninterner Selenisotopenvariationen ausreicht. Mit dem *Minimum Parameter* Kultivierungsansatz wurde eine Möglichkeit geschaffen, Selentransformationsprozesse in Pflanzen differenziert zu erforschen. Dieser Aufbau bietet zudem die Option sukzessiver Erweiterungen zur Annäherung an natürliche Bedingungen und damit zur Entwicklung einer Richtlinie für die Interpretation von Selenisotopendaten in Pflanzen aus natürlichen Systemen.

Abstract

Selenium (Se) is an essential nutrient as well as a toxin with a narrow range of tolerance. Due to heterogeneous distribution on both the local and global scale, an inadequate supply with Se is a risk factor that affects an estimated 0.5 to 1.0 billion people worldwide to varying extent. Plants are the major dietary Se source and a bottleneck for the entrance of Se into the biosphere. Redox processes thereby determine Se uptake into and distribution within the plant. Se isotope variations in geological and environmental samples proved to be a good redox tracer making them a promising tool for the exploration of the plant related Se cycle as well. However, Se isotope data are scarce and the complexity of the natural environment hardly enables systematic investigations of particular processes. This study aims to separately examine plant related processes and to figure out the relation between those and isotope signatures induced by them. To reach these goals a *Minimum Parameter* approach was chosen, in which plants were cultivated in closed and controlled systems with a minimum of external influences. Phytoagar, an artificial growth medium free of nutrients, was doped with Se in varying species and concentrations. Plants were cultivated therein in closed box systems. Mass balancing and the determination of Se isotope signatures in the compartments enabled the quantification of uptake, translocation and volatilization as well as isotope fractionation induced by them. Se isotope ratios (δ^{82}Se) were detected using hydride generation multicollector inductively coupled mass spectrometry (HG-MC-ICP-MS) ($\delta^{82}Se\ [‰] = ((^{82/76}Se_{sample})/(^{82/76}Se_{standard})-1)*1000$). This technique is characterized by high precision, but also high sensitivity on matrix effects. The gain of accurate and valid data therefore requires several steps of preceding sample treatment. A comprehensive procedure for the reasonable, precise and valid determination of stable Se isotopes in plants and phytoagar samples was developed in this study. Basis was the transformation of plants and phytoagar into a liquid form and the reduction of organic compounds, particularly organic Se species, to a minimum. Building on this, matrix compounds were separated from sample-Se in order to avoid mass interferences and inhibitions occurring within Se isotope analytics. A variety of methods for the particular steps was modified or developed with regard to the target matrices and systematically tested on their efficiency, validity and potential limitations.

A microwave digestion procedure with two reaction chambers was proved to be most suitable for plant tissue, having a mineralization rate of 99.3 (±0.4) % and reproducibly full Se recovery. For phytoagar, the developed vacuum filtration procedure was most reliable with limitations regarding organic residuals. The application of chromatographic anion exchange for matrix separation revealed a high matrix dependency, which was not detected for selective Se retention on thiol groups. Both methods were able to fully remove critical elements, but retained organic residuals in the purified phases. Se separation via hydride formation completely purified sample-Se from its organic matrix. It therefore was proved to be highly suitable for organic samples. Validation tests showed that organic residuals had a severe impact on mass bias correction and yielded invalid isotope data. Anion exchange and thiol retention did thereby not meet the demands of sample purification, whereas hydride separation produced valid and reliable results with a precision in δ^{82}Se of 0.2 (±0.2) ‰ for plant and 1.1 (±0.1) ‰ for phytoagar matrices. Validity of this method was confirmed by a certified shale reference (SGR-1). This developed and validated procedure provides a solid and reliable basis for stable Se isotope determinations in organic samples and reveals organic compounds as a limiting factor for validity.

Based on this, Se isotope variations among the compartments of the *Minimum Parameter* approach were detected. Se transfers revealed a high dependence of uptake and translocation on Se source species as well as species characteristic rate limiting steps occurring at high Se exposure. Despite of high rates, volatilization and uptake did not yield significant Se isotope fractionation. In contrast, translocation induced high fractionation (Δ^{82}Se) of +2.3 ‰ to +3.5 ‰ for selenate respectively +1.2 ‰ to +1.9 ‰ for selenite supplied plants. Higher initial Se concentration thereby corresponds to lower fractionation. These results indicate shifts in metabolic Se transformations that involve key processes for Se accumulation in particular plant parts. Knowledge of these mechanisms is crucial for the investigation and assessment of adequate human Se supply via plant foods.

This thesis provides a comprehensive and validated method for Se isotope analytics in organic samples with a precision sufficient for the detection of plant internal Se isotope variations. The *Minimum Parameter* cultivation approach offers a differentiated investigation of Se transformation processes occurring related to plants. This setup has high potentials for successive extensions to approximate natural conditions and to provide a guideline for the interpretation of Se isotope data in plants derived from natural systems.

Table of contents

List of figures

List of tables

Acknowledgements

First, I would like to thank Prof. Dr. Thomas Neumann (KIT) for supervision of my Ph.D., for being first reviewer of this work, for continuous support and helpful advice on this thesis.
I want to thank Prof. Dr. Ronny Schönberg (University of Tübingen) for uncountable hours and effort he put into the analytical setup for selenium isotope determinations in his institute, for improving my focus on details regarding analytical precision and valuable discussions, and for being second reviewer.
Special thanks to Dr. Monika Stelling, the project leader of the YIG and advisor of my Ph.D. for continuous encouragement, numerous discussions, comprehensive support and advice in and beyond scientific issues, and for proofreading.
I would like to thank the members of my PhD Commission for their support: *Prof. Dr. Thomas Neumann, Prof. Dr. Ronny Schönberg, Prof. Dr. Peter Nick, PD Dr. Stefan Norra, Prof. Dr. Nico Goldscheider.*
All colleagues from the Institute of Applied Geosciences and the Botany Institute at KIT as well as from the Isotope Geochemistry Group at University of Tübingen are acknowledged for continuous technical and scientific support, particularly

- Claudia Mößner for uncountable ICP-MS measurements with complicated sample matrices and valuable advice on method optimization
- Dr. Alexandra Nothstein for developing the concept of plant cultivation in phytoagar, for explaining and showing how to reasonably apply plant cultivation methods and for advice in many other applications in the laboratory

- Dr. Elisabeth Eiche for scientific support concerning selenium behavior in the natural and laboratory environment and for providing plant and soil samples from Punjab, India
- Dr. Stephan König for spending much time and motivation in analyzing my samples and dealing with challenging matrices, for investing time and energy in the improvement of the analytical setting and troubleshooting, for detailed explanations and demonstrations of the instrumental setup and for valuable advice on selenium isotope analytics and sample preparation
- Timon Kurzawa and Elmar Reitter for support in selenium isotope analyses and discussions
- Dr. Michael Riemann, Rita Brendel, Ronja Kammerichs and Holger Ludwig for support in the performance of plant cultivation and plant tissue and phytoagar preparation despite of a variety of special requests
- Chris Buschhaus for TOC measurements on complicated matrices

Dr. Markus Lenz and Karen Viacava (Institute for Ecopreneurship, FHNW Basel) are acknowledged for analyzing various samples on selenium species composition and for data assessment in their laboratory in Basel.
Thank you to Dr. Kathrin Schilling and Prof. Dr. Tom Johnson for inviting me to their laboratory in Urbana-Champaign in 2012 and demonstrating their methods on selenium isotope analyses and sample preparation, and for continuous advice on those issues during the last years.
I would like to thank my fellow Ph.D. students from AGW – Nicolas Börsig, Arno Hartmann, Andreas Holbach, Peter Illner, Olga Körting, Josephin Mühlbach, Alexandra Nothstein, Sebastian Potsch, Xiaohui Tang and Chen Yuen – for a nice and motivating atmosphere and for discussions on daily life issues occurring during Ph.D. time.
Thank you to Moritz Reuleaux for dedicated and detail oriented proofreading on language and style, which clearly facilitated the reader to follow my lines of argumentation, and made reading significantly less confusing.

For exchange and discussions on all the non-scientific issues occurring in academia, for several years spent on the improvement of the Ph.D. situation at KIT and beyond – with varying extent of success – for teaching me valuable lessons in diplomacy and for lots of amusing episodes I would like to thank everyone feeling associated to PaKIT.
Last but not least I would like to particularly thank my family and friends for continuous encouragement and support during the last 3.5 years, and for being unconditionally proud of me without effectively understanding why.

Funding

This work was part of Dr. Monika Stelling's Young Investigator Group (YIG), whose funding was provided by the "Concept of the Future" of KIT within the framework of the German Excellence Initiative. Travel grants for the attendance at the 3rd International Selenium Conference in Hefei, China, the Goldschmidt Conference 2015 in Prag, Czech Republic, as well as the Research Management Training Workshop in Brussels were provided by the Graduate School of Climate and Environment (GRACE), KIT.

1 Introduction

1.1 Motivation

Selenium (Se) plays a key role for human nutrition, animal health and environmental systems, as Se is an essential nutrient for all mammals and a toxin for any organism with a narrow tolerance range (Rayman, 2006). However, Se cycling in the biosphere and environment is not fully investigated yet. In terrestrial environments, plants deal as a bottleneck for Se in entering the biosphere and are therefore of particular relevance (Wu, 2004). Plant Se metabolism is characterized by Se species dependent uptake and reduction processes and the adjacent formation of particular molecules influencing accumulation or depletion in plants. As Se is a very redox sensitive element, the Se cycle in the environment and in plants is complex, transient and individual regarding ecosystem, land use and plant species (chapter 2.2).

Se stable isotope signatures proved to be a precise redox tracer and able to reconstruct Se related processes within geological and environmental samples. Spot tests on plant samples in different environments indicated high variations in Se isotope composition that were specific for the given ecosystem (chapter 2.3). Therefore, the determination of Se isotope composition of single samples as a state parameter might be able to give insight into Se related processes that took place and to reconstruct how the current Se status came to be. Therefore it is a promising and feasible tool to explore the Se cycle in plants growing under various conditions and reveal the individual causes of Se related issues.

A prerequisite is the full understanding of relations between characteristic Se isotope signatures in plants and their underlying processes, which are not systematically investigated yet. This requires the possibility to separately determine plant related Se processes that are naturally occurring parallel to each other. Furthermore, particular sample preparation and analytical methods have to be developed, implemented and validated to gain reliable Se isotope data from plants. With both issues requiring systematic pre-studies their combination contain a considerable potential to provide a solid and reliable basis for the use of Se isotope signatures in plants and other biological samples.

1.2 Goals and objectives

The *main goal* of this study is the provision of a methodical basis to separately investigate Se transformation processes that in nature occurring in parallel (e.g. inorganic and microbial Se reduction in soil and water, uptake of different Se species) and the extent of Se isotope variations induced by those. Cultivation experiments in the laboratory are a reasonable approach as they facilitate closed and controlled conditions as well as a continuous monitoring and reproduction. Independent, parallel setups enable the separate supplementation of Se species and concentrations. Therefore the *first objective* of this thesis is the development of a *Minimum Parameter* approach, which allows focusing on single processes only including H_2O, Se and the plants, avoiding the influence of soil components and macronutrients. This includes the adaption of existing concepts to the particular demands of Se isotope analytics, the monitoring of parameters and their stability as well as the examination of reasonable Se species and concentration supply with regard to Se isotope variations among compartments of the closed system.

Quantification of Se transfer processes and Se isotope fractionations presupposes the determinations of Se content and Se isotope ratio in any compartment. Precise and valid Se isotope analytics requires preceding sample treatment and selective purification from matrices, which is particularly challenging with organic rich samples deriving from the *Minimum Parameter* cultivation. Therefore, the *second objective* of this study is the development and implementation of comprehensive procedures meeting the analytical challenges. This includes the development of new methods or the adaption of existing ones to particular sample demands as well as the systematic monitoring of efficiency with particular regard to analytical validation.

2 State of the art

2.1 Selenium – properties and relevance for environment and health

2.1.1 Chemical and geochemical properties

Selenium is an element in the periodic table with number 34 in group 16, period 4 and an atomic mass of 78.96 amu (Wieser et al., 2013) (Figure 1 (a)). It belongs to the group of chalcophile elements and shares particular properties with other elements of this group such as sulphur (S) and tellurium (Te). Se tends to form insoluble complexes and to be incorporated into crystal gratings of sulphidic minerals, substituting S. Additionally it forms mobile oxyanions and a variation of organic complexes (Wiberg et al., 2001; Lenz and Lens, 2009). Redox and pH conditions are an important factor for the stability of or transformation into a species or compound, particularly with Se, which has a high sensitivity on pH and redox changes (Takeno, 2005) (Figure 1 (b)).

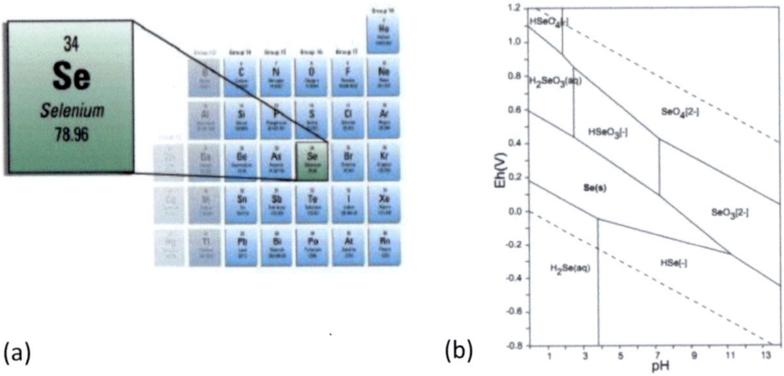

Figure 1(a) The position of Se within the periodic table of elements (USGS, 2015), (b) Pourbaix diagram of Se (Se-O-H availability, 20°C, 1M Se, PhreeqC based model) (Takeno, 2005).

In oxic environments Se tends to form selenate anions (SeO_4^{2-}) that are thermodynamically favored in stability and characterized by their high solubility and mobility. Selenate has a low affinity to reactions due to its tetraedic structure being energetically advantageous and therefore very stable (Olin et al., 2005). Oxygen exchange rates with H_2O were reported to be exceptionally low with a half-life of 10^6 years at 25°C and neutral pH conditions (Kaneko and Poulson, 2012). Selenate forms either bidentate outerpheric (Figure 2 (a)) or monodentate innerspheric (Figure 2 (b)) complexes, e.g. on ferric (Fe(III)) oxide and hydroxide surfaces. These sorption mechanisms are relatively weak and reversible (Su and Suarez, 2000; Peak and Sparks, 2002).

The oxyanion selenite (SeO_3^{2-}), available at moderately oxic conditions within the entire pH range, is soluble and mobile as well, but more affine to sorption (e.g. to iron oxides, clay minerals) as innerspheric bidentate complexes, which are more stable and rather irreversible (Zhang and Sparks, 1990; Su and Suarez, 2000) (Figure 2 (c)). Furthermore selenite tends to reduction processes to Se(0) or Se(-II) and, as a consequence, the incorporation into mineral or organic particles. Oxygen exchange with H_2O is very quick and frequent (Kaneko and Poulson, 2012).

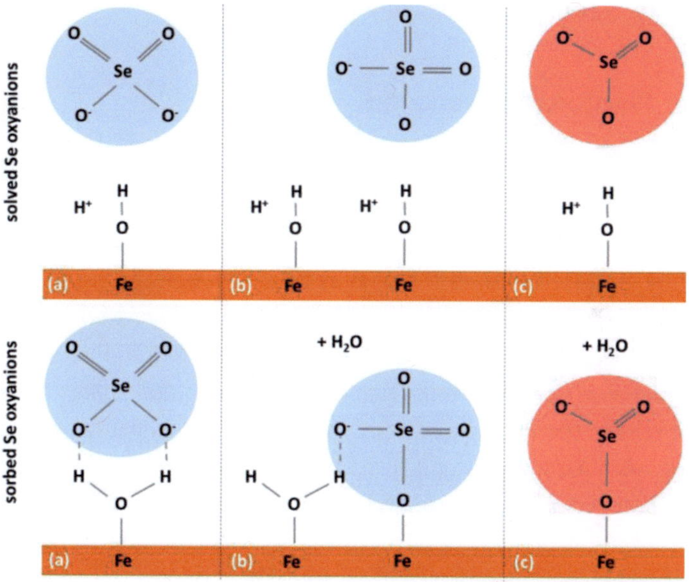

Figure 2: Simplified scheme of sorption mechanisms on ferric hydroxides of the Se oxyanions selenate (blue) and selenite (red) – (a) selenate sorption as bidentate outerspheric complex, (b) selenate sorption as monodentate innerspheric complex, (c) selenite as bidentate innerspheric complex (data from Zhang and Sparks, (1990), Su and Suarez (2000), Peak and Sparks (2002)).

Elemental Se (Se(0)) is hardly soluble and tends to precipitation on mineral surfaces or as nanoparticles. Being different from the other Se species, Se(0) plays a minor role in environmental and biological processes, although Se(0) can be bioavailable to particular organisms and even produced from Se oxyanions by microbials under aerobic and anaerobic conditions (Winkel et al., 2012; Jain et al., 2014).

Se's fully reduced oxidation state, Se(-II), is very reactive and easily incorporates into mineral complexes and organic molecules. Under strongly reducing and acidic conditions it forms gaseous hydrides (H_2Se), which is a characteristic property for Se and only shared by few other elements (e.g. Ge, As, Br, Te, Sb, Bi). The use of this characteristic for analytical purposes is widely applied (e.g. Ribeiro et al., 2004; Rouxel et al., 2004).

2.1.2 Physiological function

Se is of high environmental significance, because it is an essential nutrient and a toxin for all mammals including human beings (Rayman, 2006). It replaces S by Se in the amino acids methionine and cysteine and therefore forms the different essential amino acids selenomethionine (SeMet) and selenocysteine. These are part of a variety of proteins such as thyroid hormones, enzymes protecting cells from oxidation and free radicals as well as muscle tissue and brain cells (Holben and Smith, 1999; Pillai et al., 2014). Se strengthens the immune system (Rayman, 2006; Brinkman et al., 2006) and detoxifies As(III) and Hg(II) by forming covalent As-Se and Hg-Se bonds (Gailer, 2007; Ralston and Raymond, 2013; Pickering et al., 2014). Clark et al. (1996) reported a preventive effect of Se on cancer, but the universality, the actual causes and the applicability are still controversial, whereby the medical research is very active regarding this issue (Ip, 1998; Marshall, 2014). Several studies reported furthermore the key role of Se in the Human Immunodeficiency Virus (HIV) metabolism, while stating that an adequate Se level retarded the onset of Acquired Immune Deficiency Syndrome (AIDS) and reduced the mortality of the HIV infected patients studied, although the reasons are not fully discovered yet (Semba and Grey, 2001; Baum et al., 2001, Kupka et al., 2004; Sudfeld et al., 2014 and others). On the other hand the replacement of S in amino acid in high amounts can have severe impacts on the functionality of S requiring proteins such as DNA reparing enzymes, tissue structures and functions and neural cells (Moreno-Reyes et al., 1998). The tolerance range of chronic daily uptake for human beings is on average between 40 and 400 µg, depending on age, weight and gender, whereas the recommended minimum uptake is 70 µg (Figure 3).

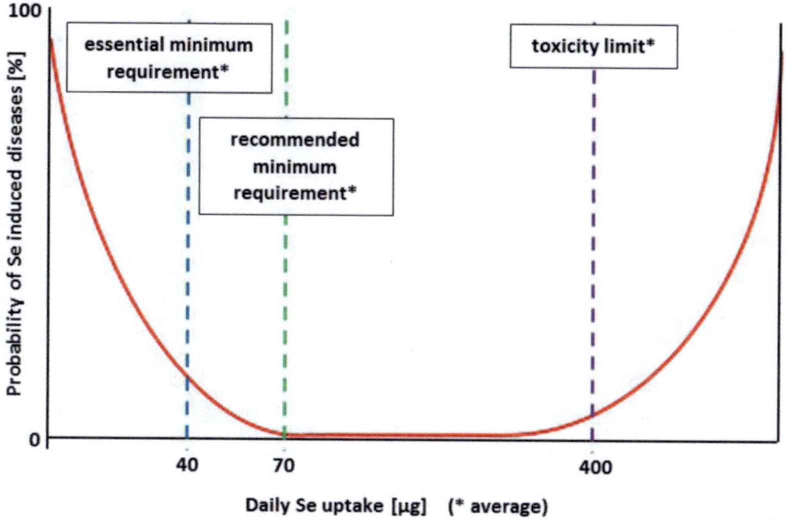

Figure 3: Relation between daily Se uptake and Se related diseases – thresholds for an average grown-up person (qualitative image) (data from Moreno-Reyes et al. (1998), Levander and Burk (2006) and Stranges et al. (2010)).

Characteristic Se toxicity symptoms are brittle nails respectively horn and hooves of livestock, hair loss and skin lesions due to a dysfunctional beta-sheet structure in those tissues (Moreno-Reyes et al., 1998; Holben and Smith, 1999) (Figure 4 (a)-(c)). Another characteristic disease is blind staggers, a neurological dysfunction particularly affecting livestock, indicated by an unsteady staggering gait and loss of vision (Moreno-Reyes et al., 1998) (Figure 4 (d)). In plants, excessive uptake might lead to reduced growth and crop failures (Dhillon et al., 2005).

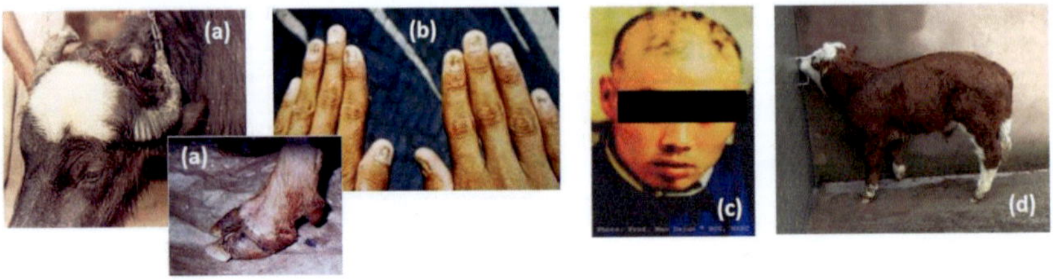

Figure 4: Se toxicity symptoms – (a) brittle horn and hooves (www.upei.ca (10.03.2014)), (b) brittle nails (Dhillon et al., 2005), (c) hair loss (Fordyce, 2007), (d) blind staggers (www.nature.com (10.03.2014)).

The quantitatively more important issue is Se deficiency. An estimated 0.5 to 1 billion people worldwide are affected from it to a varying extent (Haug et al., 2007). The Keshan disease, discovered in the Chinese province of Keshan and prevailing in China, is a severe consequence of Se deficiency. It is a cardiomyopathy and affects the functionality of the heart muscle (Stone, 2009). Another disease frequently occurring in China is the Kashin-Back disease, named after its discoverers. This type of osteoarthritis particularly affects children and their bone growth, leading to stunted development of the skeleton (Moreno-Reyes et al., 2001) (Figure 5 (a)). Another dysfunction caused by Se deficiency is the White Muscle Disease, which impacts the development of muscles and typically occurs with livestock. It usually affects the leg muscles first, but may expand to any muscle including the heart (Gunes et al., 2010) (Figure 5 (b)). Furthermore, thyroid dysfunctions as well as an increased risk of diabetes were reported to be characteristic consequences of chronic Se deficiency in human beings (Contempre et al., 1992; Stranges et al., 2010).

Aside of severe consequences for humans' and animals' health, inadequate Se supply might also cause economic losses in agriculture and livestock farming as well as threatening food security in particular regions.

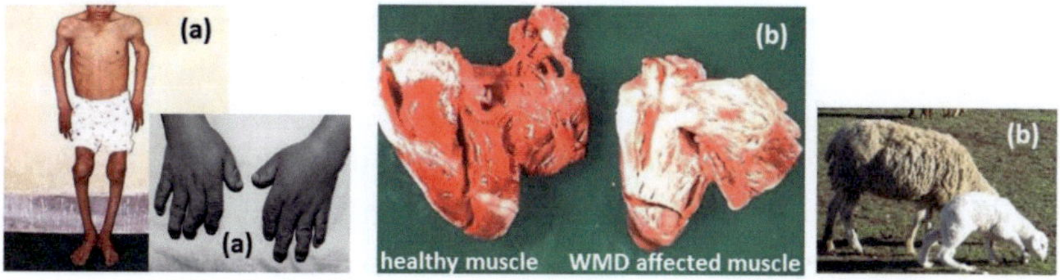

Figure 5: Se deficiency symptoms – (a) Kashin-Back disease, (b) White Muscle Disease (WMD) (http://drainameducci.blogspot.de, www.upei.ca, www.goatbiology.org (10.03.2014)).

2.1.3 Economic relevance and industrial applications

Exploiting interest and public awareness for Se and its physiological functions (chapter 2.1.2), there is a particular industry branch providing Se supplements and Se fortified food. Public health campaigns address the importance of an adequate Se supply as well. Examples of both are given in Figure 6.

Figure 6: (a) Advertisement for Se fortified potatoes *Selenella* from Italia (www.salvomessina.com (21.07.2015)), (b) naturally Se enriched potatoes *Selena* from a Se rich area in Ireland (www.selena.ie (21.07.2015), (c) Se fortified eggs (www.ocado.com (21.07.2015), (d) health campaign to promote naturally enriched garlic as dietary Se source (www.mhlw.go.jp (21.07.2015)).

Furthermore Se is of high relevance for technical and pharmaceutical applications. Figure 7 shows the industry branches using Se and their fraction of total Se used in industry as well as examples of products containing Se as a key component. For instance, dietary supplementations play an important role for humans and livestock. The semiconducting property of Se is used in electronics and energy production. In glassware, pigments are made from Se to obtain green color. Furthermore Se is applied in metalworking, among other functions as alloy component (www.selenium.de (2011)).

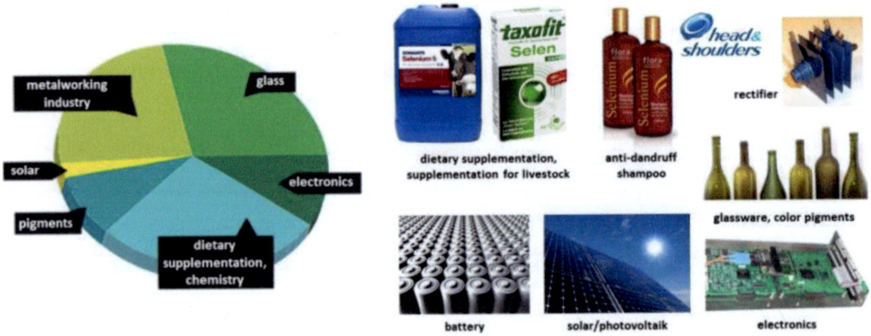

Figure 7: Industrially used Se divided into branches (left), examples of products containing Se (right) ((www.selenium.de – data from 2011) www.bembu.com, www.selenium.de (28.01.2014)).

As there are no particular Se deposits, Se is gained as a byproduct of copper (Cu) mining. Due to the expansion of the renewable energy sector, Moss et al. (2011) expect a global rise in Se demand for photovoltaic systems within the next decades. For that reason the necessity to carefully manage the scarce resource will rise, including the implementation of innovative technology in Se acquisition and recycling in order to sustainably fulfil the global Se demands (Haug et al., 2007).

2.1.4 Distribution in the environment

The relative abundance of Se within the earth crust (<16 km) is $8*10^{-5}$ % (Bodik et al., 1988) making Se a trace element. Soil contents are on global average 0.05 ppm and in most waters <0.1 µg L^{-1}. However, due to Se's redox sensitivity on environmental conditions its distribution is heterogeneous especially on the earth surface, which causes a high range of soil Se concentrations with hot spots of up to 1200 ppm (Dhillon and Dhillon, 2003; Fernández-Martínez and Charlet, 2009). The near surface environment shows a very dynamic Se cycle characterized by redox changes (Figure 8). This applies especially for the *Critical Zone*, which is characterized by the interaction of lithosphere, pedosphere, hydrosphere, biosphere and atmosphere (US NRC, 2001).

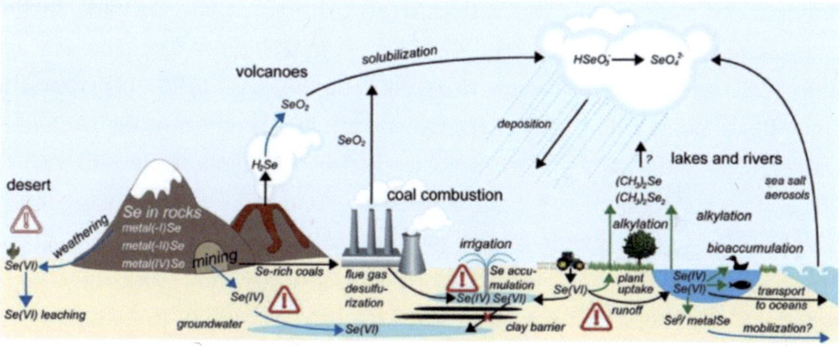

Figure 8: Schematic global cycle of Se with main focus on the terrestrial environment. Blue arrows – process involves oxidation of Se species, green arrows – process involves reduction of Se species. Warning symbols indicate specific environmental settings that are at risk of either developing Se deficiency (open warning symbol) or Se excess (shaded warning symbol) (Winkel et al., 2012).

Potential geogenic Se sources are generally organic and S rich source rocks such as black shales, carbonaceous limestones, carbonaceaous cherts, mudstones and seleniferous coal. Main anthropogenic sources are fossil fuel combustion and sulphidic ore mining. By exposing Se containing ores to the surface, Se might be mobilized through oxidation (Wang and Gao, 2001; Wen and Qiu, 2002; Zhu et al., 2008a). Mobilized Se is distributed via the atmo- and hydrosphere. Potential sinks are lakes, organic and iron rich soils as well as the biosphere in general. When available, Se tends to excessive bioaccumulation and –magnification and therefore high enrichment in organisms (Wu 2004; Winkel et al., 2012). One example for Se excess is the San Joaquin Valley in California, USA. The area is characterized by marine shales as Se rich source rocks and subsequently by Se rich groundwater in adjacent aquifers. Intensive agricultural activities including continuous excessive irrigation with Se rich water from deeper aquifers have led to Se enrichment in soils and plants. Due to consistently low permeable shale layers, Se enriched seepage water can hardly be naturally discharged or diluted. Within the last decades, high amounts of this water were artificially drained and passed into the wetlands of Kesterson National Park. In this area, Se concentrations in water were reported to be 300 µg L^{-1} on average and 4200 µg L^{-1} on maximum measured in the late 1980s (Presser and Ohlendorf, 1987; Fan et al., 1988; Ohlendorf, 2002) and thereby exceeded the WHO drinking water threshold by factor 30 respectively 420 (WHO, 2011). The consequences for the local ecosystem were devastating. Ohlendorf (1986) estimated that 20% of all birds had deformities and 40% of the embryos died before hatching. Deformities and high mortality rates were reported for several fish species as well (Ohlendorf, 2002). In 1986, Kesterson National Park was officially declared as a waste dump and the drain into San Francisco Bay (Pacific Ocean) was closed. However, significant amounts of highly Se contaminated water are still draining into the Pacific Ocean (Ohlendorf, 2002). Since the 1980s until today Kesterson National Park was an object of studies on potential phytoremediation measures concerning Se contaminated ecosystems (e.g. Banuelos and Lin, 2005).

Another example for critical Se accumulation is a region in Eastern Punjab, India, that is characterized by intensive agricultural land use as well. The agricultural land is continuously irrigated with Se rich groundwater having concentrations of average 69.5 µg L^{-1} and maximum 341 µg L^{-1}. Consequences were the Se enrichment in soils (average 6.5 ppm) and plants (3 – 670 ppm) (Bajaj et al., 2011) clearly exceeding the US EPA threshold of 5 ppm Se for forage (US-EPA, 2000). Crop failures and economic damages could be observed alongside Se toxicity symptoms in the local population (Dhillon et al., 2005). The issue escalated with the change of wheat-maize to wheat-rice rotating cultivation, probably because rice cultivation requires higher amounts of irrigation water and influences the Se retention and mobility with a changing redox environment (Bajaj et al., 2011). Both examples show that the accumulation of Se depends on various factors like source rock Se concentration, redox environment, land use and water management. Furthermore the mineral composition of the soil, the plant species and its affinity to Se as well as climate and hydrological conditions might advance Se accumulation (Winkel et al., 2012).

In areas with low retention potential, for instance caused by low organic content and low sorption potential due to lack of clay minerals and iron oxides, Se tends to be leached out and depleted. In nature, this often occurs simultaneously with low water retention capacity and oxic conditions, which both increasing the mobility of Se (Figure 9). On the contrary, high Se retention potentials in soils with high organic, iron and clay mineral content might immobilize Se in the soil and prevent plant uptake. Furthermore a high water retention capacity might induce stagnating water. Combined with high organic content reducing conditions might arise, which tend to transform Se into the insoluble and hardly bioavailable Se(0) (Sarret et al., 2005; Chakraborty et al., 2010). Wang and Gao (2001) reported a clear coincidence of different factors negatively influencing the bioavailability of Se – low soil Se concentrations, high evaporation rates, slightly acidic pH values and high soil organic content – with the prevalence of Se deficiency disease patterns in China. Particularly in North-Eastern China as well as in parts of Central and South China the local population is affected by Keshan disease, Kashin-Back disease or both. Se deficiency was reported in huge parts of sub-Saharan Africa by Chilimba et al. (2011) and Joy et al. (2014), especially Malawi and Sambia. A combination of low soil Se, low soil pH and high retention potential by mineral composition caused low Se content in local staple food and Se deficiency symptoms in the local population. Due to Se poor source rocks, parts of North-Western and Eastern USA as well as huge parts of Europe are depleted in Se, making Se supplementation necessary particularly for cattle, cows, sheep and horses (Oldfield, 2002) and led to field experiments on Se biofortification on national scale, e.g. in Finland (Alfthan et al., 2015). Figure 9 schematically shows the influence factors and their interaction in Se depletion and accumulation potentially leading to Se deficiency or toxicity.

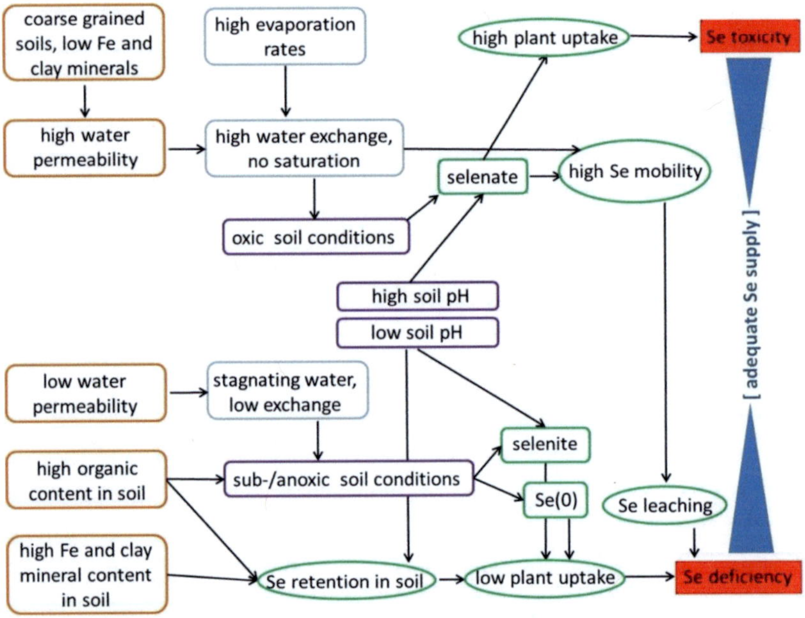

Figure 9: Scheme of influence factors for Se mobility and the origins of Se related problems (simplified) (circled blue – hydrological conditions, brown – geological and soil properties, purple – pH and redox conditions, green – Se related processes, red – Se related problems).

Several factors indicate an increase in Se related problems within the next decades. Climate change will probably raise precipitation variabilities and enhance local contrasts in precipitation amounts (IPCC, 2013), which in turn will influence redox conditions, leaching and the composition of soils. Population and space pressure as well as the rising popularity of land as investment possibility (*land grabbing*) will expand the area of intensively cultivated agricultural land (Taagepera, 2014; Antonelli et al., 2015), which is a critical issue in context of Se as exemplarily reported for San Joaquin and Punjab. Due to economic growth particularly in emerging countries, the demand of metals and energy and therefore mining activities will probably grow (Legarth, 1996; Miller, 2013) and thereby extent the main anthropogenic Se source. That is why the knowledge on environmental Se cycling and the elaboration of solution concepts for Se deficiency and toxicity will gain importance.

An overview about the global Se distribution is given by Oldfield (2002). A detailed review on the global Se cycle is given by Winkel et al. (2012).

2.2 Selenium and plants

2.2.1 Role of plants

As described in chapter 2.1.4, Se is a biophilic element, which tends to accumulate in the biosphere and to magnify along the food chain (Wu, 2004). Plants are bottlenecks for the entrance of Se into the biosphere (Figure 10) and therefore key objectives for the investigation of Se pathways and management of adequate Se supply of humans and animals. Plant foods are the major dietary Se source in most countries, followed by meat and fish. Drinking water plays a minor role (US-NIH, 2013).

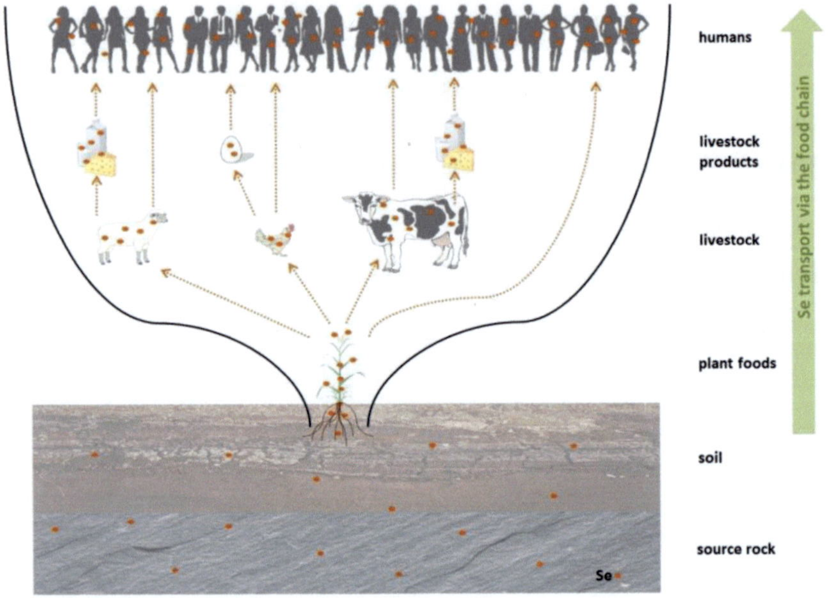

Figure 10: Simplified scheme of Se entering and transport into the food chain via plants and bioaccumulation along the food chain.

Various parameters influence the uptake of Se into plants, the soil concentration is only one of them (Ellis and Salt, 2003; Winkel et al., 2012) (Figure 9). Se underlies active uptake mechanisms by nutrient transporters and transformation processes such as reduction and incorporation within the plants. The role of Se in plants is not fully investigated yet. It is probably not essential, but might fulfill several functions such as protection against certain (a)biotic stresses in moderate concentrations (Hasanuzzaman et al., 2014), but leads to phytotoxicity in high concentrations (Dhillon et al., 2005). Plants take up Se via macronutrient transporters and include it into their tissue or emit it to the atmosphere (Li et al., 2008). The oxidation state of Se plays a major role in forming bindings and species, and those determine uptake, translocation and accumulation pathways. Furthermore, the role of the plant species is important, because those significantly differ regarding uptake, accumulation and volatilization affinity. This chapter presents the common sense of Se transport and transformation pathways related to plants and regarding various influence factors.

2.2.2 Se uptake and translocation processes

The processes of Se uptake and translocation by plants dependent on source species were investigated by several studies using cultivation experiments. Many of them focussed on Se non-accumulator plants that are frequent as agricultural crops, e.g. maize, rice and wheat. As they are frequent in nature and most relevant for plant uptake, the Se oxyanions selenate and selenite as well as the organic species and amino acid SeMet were regarded. Table 1 shows experimental parameters and translocation factors (T) of the studies mentioned. The translocation factors were calculated via the absolute Se amounts measured in the plant parts (a(Se) [μg]) according to Equation (1).

$$T = \frac{a(Se)_{shoot}}{a(Se)_{root}} \quad (1)$$

Table 1: Overview over selected studies on cultivation of Se non-accumulator plants with varying Se concentrations, growth media and cultivation times and their results concerning translocation (translocation factor according to Equation (1)) (volatilization was neglected in all studies)

	Kikkert and Berkelaar (2013)	Nothstein (2015)	Longchamp et al. (2015)
cultivated plant	wheat	rice	maize
Se supplied (µg L^{-1})	40 - 400	250	1000
growth medium	soil + optimum nutrient supply	phytoagar + no nutrients	optimum nutrient solution
cultivation time	300 min (5 h)	16 days (384 h)	until maturity (~6 mon) (>4000 h)
Translocation factor			
selenate supply	>1	4.6	13.3
selenite supply	<0.1	0.5	0.6
SeMet supply	0.25 - 0.75	N/A	N/A

Figure 11 schematically shows the absolute Se uptake and translocation depending on the source species as well as the average species distribution in roots and shoots. Thereby, selenate and selenite illustrations contain average values of Kikkert and Berkelaar (2013), Nothstein (2015) and Longchamp et al. (2015) and the SeMet image is based on Kikkert and Berkelaar (2013) as the only study investigated SeMet uptake. The Se species distribution within the plants derives from Kahakachchi et al. (2004). The figure only gives a qualitative impression as the studies differ in framework conditions such as growth medium, Se concentration supplied, cultivation time and plant species. Nevertheless tendencies are recognizable that correspond with previous studies (e.g. Zayed et al. (1998), Li et al. (2008), De Souza et al. (1998)).

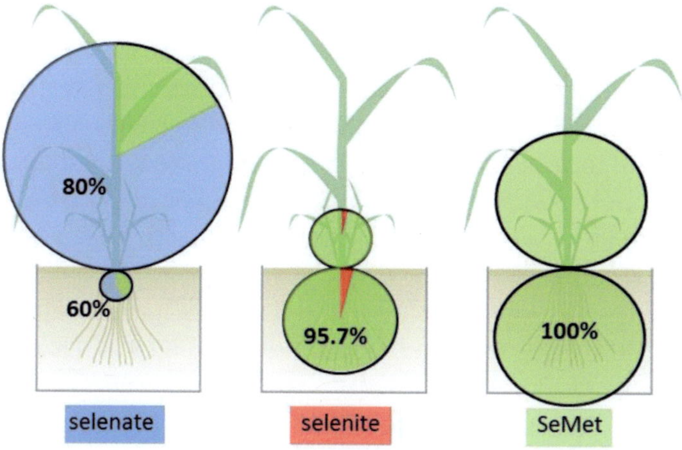

Figure 11: Schemes of Se distribution and species abundances in different plant parts according to Longchamp et al. (2015), Kikkert and Berkelaar (2013), Nothstein (2015) (wheat, maize, rice) and Kahakachchi et al. (2004) (*Brassica*) (Table 1). The blue circle parts represent the selenate fraction, the red ones the selenite fraction, the green ones organic Se species, mainly SeMet.

All presented studies agreed that selenate had the highest uptake and translocation among all supplied Se species (Longchamp et al. (2015), Kikkert and Berkelaar (2013), Nothstein (2015), Zayed et al. (1998), Li et al. (2008), De Souza et al. (1998)). Reasons are on the one hand the high amount of sulphate transporters via which selenate is taken up (Li et al., 2008), on the other hand the high thermodynamic stability of selenate and therefore low affinity to species transformation or retention (chapter 2.1.1 and 2.1.4). As a consequence, selenate is very affine to enter the plant and being translocated within it. The species distribution in plants, reported by Kahakachchi et al. (2004), shows that the very dominant fraction was still available as selenate, which applies even more for shoots than for roots (Figure 11 left). This confirms the idea of Gissel-Nielson (1984) that selenate mainly remains in this species while transported to the shoots whereas Se taken up as selenite is quickly transformed and transported as organically bound Se. That the entire residual Se could be identified as organic Se compounds, mainly SeMet (Kahakachchi et al. (2004)), indicates that selenite, which is a mandatory intermediate in the reduction and transformation of selenate to organic Se species, is indeed very unstable within the plant and affine to reduction and organic incorporation. The selenite supplementation setups confirm this assumption as almost 96% of the selenite taken up was found as organically bound Se in the plants (Figure 11 center) (Kahakachchi et al. (2004)). Additionally, the uptake rate of selenite was - by far - lower than this of selenate or SeMet, and the translocation rate was very low in relation to selenate and SeMet (Longchamp et al. (2015), Kikkert and Berkelaar (2013), Nothstein (2015), Zayed et al. (1998), Li et al. (2008), De Souza et al. (1998)). This might depend on the uptake mechanism, which is not as clearly identified as the one of selenate. Li et al. (2008) proved that it was an active uptake process and they assumed that it was probably taken up by phosphate transporters due to their abundance and the chemical similarity of selenite and phosphate, whereas Zhao et al. (2010) hypothesized that it might be taken up by silicon transporters. However, the more probable reason for the lower uptake and translocation rates is the quick transformation of selenite into organic Se, its incorporation into the plant tissue (Figure 11 center) (Kahakachchi et al. (2004); De Souza et al. (1998); Li et al. (2008)) and therefore the retention of the major fraction within the roots. The small fraction translocated was mainly transformed into organic species before translocation (Gissel-Nielson (1984); Kahakachchi et al. (2004); De Souza et al. (1998)). Nothstein (2015) however reported that selenate supplied rice plants stored 38-54% in organic form with minor differences between roots and shoots, whereas selenite supplemented rice plants stored 85-100% as Se_{org} in the roots and 64-80% in the shoots respectively.

Data on the transport pathways of Se in plants supplied with organic species is sparse although organic Se is frequent in the environment and almost 100% available as SeMet there and within non-accumulator plants (Li et al., 2008, Neal 1995). According to Sandholm et al. (1973) SeMet is taken up via amino acid transporters and therefore only competes with the amino acid methionine. In short term experiments (300 min exposure) with wheat the uptake rates were detected 40-100 times higher than for anorganic Se species and the shoot concentrations were highest, too. In contrast to selenate and selenite supplied plants, SeMet was evenly distributed between roots and shoots (Kikkert and Berkelaar, 2013). The uptake rates as well as the bioavailability and the affinity for incorporation are very high, which leads to both – incorporation into the roots and translocation followed by incorporation into the shoots.

The tendencies are similar in those three studies, although they use different growth media, cultivation times, supplied Se concentrations and plant species. The translocation factor of selenate increases with extended cultivation time and Se concentration, which might be a result of higher phytomass growth rates in shoots compared to roots and with limited interaction of selenate with the root tissue. Selenite was generally mainly stored in the roots and hardly translocated, but with increasing cultivation time and Se concentration the translocation factor rises probably due to rate limiting steps for reduction and incorporation and therefore backlog effects.

On the molecular level, Mounicou et al. (2006) reported that the roots contained mainly high molecular weight Se molecules, which are less mobile and mainly the product of Se incorporation into the plant tissue or formation of proteins, whereas the leaves contained low molecular weight Se molecules, which are the more mobile inorganic species, especially selenate, as well as small and soluble organic compounds such as SeMet that were not incorporated yet and might partly be transformed to volatile species such as methylselenides (chapter 2.2.4).

2.2.3 Se accumulating plants

When characterizing Se uptake and accumulation as well as the accompanying metabolic pathways, three types of plants are commonly differentiated: the non-accumulators, the secondary accumulators and the hyperaccumulators. Concerning Se species distribution, the main difference is the varying character and composition of organic Se compounds.

Chapter 2.2.1 deals with non-accumulating plants, which are most frequent and most important for agriculture, e.g. rice, maize and wheat. They accumulate the biophilic Se in moderate amounts compared to the other plant types. As selenite and the minor fraction of selenate is directly transformed into the amino acid SeMet and then included into proteins, effectively the entire organic Se is available as solved SeMet or, in lower amounts, $SeCys_2$, as water soluble non protein forms (e.g. methylselenomethionine) or as protein part, with a clear dominance of SeMet (Mounicou et al., 2006, Neal 1995). Mounicou et al. (2006) additionally reported that non-accumulator plants contained higher ratios of Se bearing proteins than secondary or hyperaccumulating plants, probably due to the lack of Se exclusion mechanisms from plant metabolism.

Secondary accumulators, e.g. *Brassica juncea*, accumulate high amounts of Se if present, but they do not have competitive advantage on Se rich sites and lacking Se does not impair their growth and development (USDA, 2014). Secondary accumulators contain a variety of organic Se species. The amino acids SeMet and $SeCys_2$ were found in similar proportions (Mounicou et al., 2006). Moreover, up to 50% of the total organic Se species were characterized as water soluble non protein forms, mainly methylselenomethionine. Other studies assumed that the transformation of SeMet to methylselenomethionine and the further transformation to the volatile species dimethylselenide (DMSe) was an active mechanism to exclude Se from the plant tissue and therefore enhance the Se tolerance (Neal 1995, Tagmount et al. 2002, Goa et al. 2000).

Hyperaccumulators (or primary accumulators), e.g. *Astralagus bisulcatus, Stanleya pinnata*, accumulate exceptionally high amounts of Se. By early Se accumulation and due to particular strategies for tolerating high Se amounts, they are perfectly adapted to Se rich areas and more competitive than Se sensitive plants. Additionally a herbicide effect due to high Se emissions was reported (El Mehdawi et al., 2011). In hyperaccumulator plants, selenite is directly transformed to methylselenocysteine instead of SeMet. Methylselenocysteine cannot be incorporated into the plant tissue. It is directly transformed into the volatile dimethyldiselenide (DMDSe) and emitted into the atmosphere (*volatilization*) (LeDuc et al. (2004)). DMDSe contains two Se atoms, which ensures a more efficient Se emission than the formation of DMSe by secondary accumulators. However, both accumulator types are characterized by active Se exclusion mechanisms based on volatilization. Active mechanisms are generally energy-intensive, and volatilization prevents material and energy recovery, which is why plants will only use them if necessary. Therefore it is probable that those mechanisms are activated and controlled by Se concentration thresholds and accumulating plants might additionally be able to conduct non-accumulators' Se pathways if Se is only present in moderate amounts. Detailed investigations on this issue – beside the ones mentioned – are lacking yet.

However, the plant species and its affiliation to a particular accumulator type are of high importance not only for uptake rates, but also for the Se metabolism and the underlying reactions within and related to the plant. Figure 12 summarizes the characteristic metabolic pathways regarding the accumulator type.

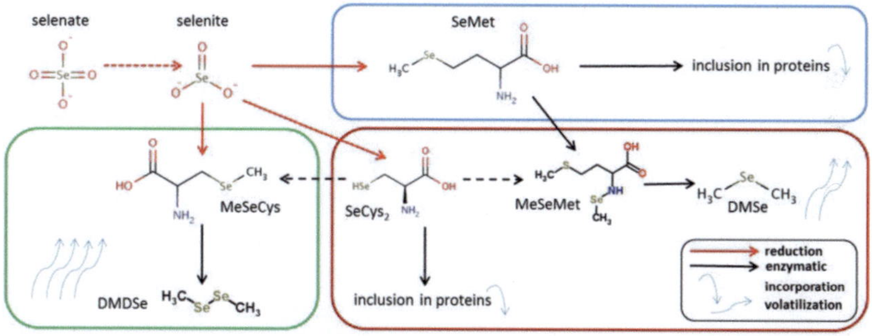

Figure 12: Simplified scheme of Se transformation processes within plants characteristic for (but not limited to) the group of non-accumulators (blue), secondary accumulators (red) and hyperaccumulators (green) (DMDSe – dimethyldiselenide, DMSe – dimethylselenide, MeSeMet – methylselenomethionine, MeSeCys – methylselenocysteine, SeMet – selenomethionine, $SeCys_2$ – selenocysteine) (data from Neal (1995), De Souza et al. (1998), Goa et al. (2000), Tagmount et al. (2002), LeDuc et al. (2004), Mounicou et al. (2006) and Jones et al. (2014)).

2.2.4 Se volatilization

Se volatilization is defined as biologically induced emission of Se as gaseous component into the atmosphere. It is a very important process within the plant related Se cycle, quantitatively and regarding ability and tolerance for Se accumulation. As discussed in chapter 2.2.3, the transformation of Se into volatile species is an active mechanism to tolerate harmful Se stress. The volatilization rates can reach up to 77% of the taken up Se (Jones et al., 2014). However, the amount is very much dependent on Se species and supplied concentration as well as the plant species and its particular Se metabolism. Because of the necessity of complex setups experimental data on Se volatilization in plants is rare. Terry et al. (1992) determined the volatilization rates of a variety of plant species. In general, they found a high correlation between tissue concentration and volatilization rate, both in variations of factor 10 (*Se volatilized per phytomass*) respectively 23 (*Se volatilized per leaf area*). This confirms the results presented in chapter 2.2.3 that the affinity to accumulate Se is directly connected to the tendency of Se emission via volatilization. Among broccoli and cabbage, *Oryza sativa* (rice) was reported to have the highest volatilization rates of up to 350 µg Se m^{-2} (*leaf area*) d^{-1} respectively 2500 µg Se kg^{-1} (*phytomass dry weight*) d^{-1}. The volatilization rates per leaf area are exceptionally high for rice, whereas the rates per phytomass are intermediate. Also more than 70% of the Se taken up in rice was stored in the leaves, which is the highest value of all plant species examined. This indicates that the dominant Se fraction volatilized was emitted via the leaves. De Souza et al. (1998) examined the volatilization rates for *Brassica juncea* supplied with selenate and selenite in parallel setups. Volatilization was 2-3 times higher with selenite supplied plants, whereby the accumulation and volatilization rates increased linearly with source concentration in both setups. It was assumed that the volatilization from selenate was limited by selenate reduction, which is thermodynamic-cally unfavored and therefore works as a limiting factor. Volatilization of selenite supplied plants may be limited by selenite uptake and by conversion of SeMet to DMSe, probably not by selenite reduction and transformation to

SeMet based on the observed instability of selenite in plants (chapter 2.2.2) (De Souza et al. 1998).
Jones et al. (2014) investigated the sources of volatilization with different accumulator plants and found that the volatilization rate in hyperaccumulator plants (77%) was elevated by factor 3 compared to a secondary accumulator (25%) at equal Se amounts supplied. Furthermore they discovered that in the hyperaccumulator setup the very dominant fraction was volatilized via the shoots, whereas the fractions volatilized from the secondary accumulator setup were similarly divided between shoots and roots (56%, 44%). It was assumed that $SeCys_2$, a selenite metabolite from the hyperaccumulators was more mobile and reached the shoots in higher amounts than SeMet, the analogous product in the secondary accumulator Se metabolism (Figure 12). SeMet was tendentially retained in the roots and partly incorporated, partly transformed into volatile species there (Jones et al., 2014).

2.3 The stable selenium isotope system

Stable isotopes are atoms with the same number of protons and electrons, but a differing number of neutrons. They share the same space in the periodic table of elements (*iso* (greek) – same, *topos* (greek) – place), but have little variations in their physico-chemical behavior, especially at chemical reactions or phase transitions (Hoefs, 2009).
The partitioning of Se isotopes between substances or phases is called isotope fractionation. The major processes inducing isotope fractionations are isotope exchange reactions (equilibrium isotope distribution) and kinetic processes. Furthermore there are mass independent processes inducing isotope fractionations, which are neglected in this study. Details are given by Hoefs (2009).
Isotope exchange summarizes all processes with no chemical net reaction, but changes in isotope distribution between substances, phases or single molecules in order to reach an isotopic equilibrium. Isotopic exchange reactions can be described as special cases of chemical equilibria as in

Equation (2), characterized by the equilibrium constant K defined in Equation (3). A and B are different substances, phases or molecules, a and b are abundances and 1 and 2 represent the lighter respectively heavier isotope.

$$a * A_1 + b * B_2 = a * A_2 + b * B_1 \tag{2}$$

$$K = \frac{(\frac{A_2}{A_1})^a}{(\frac{B_2}{B_1})^b} \tag{3}$$

Molecular bindings are influenced by rotational, translational and vibrational energy. As rotation and translation do not significantly differ between the compartments, differences in vibrational energy are the predominant cause for isotope fractionation by isotope exchange. As a consequence, the binding energy is higher in molecules containing the heavy isotope instead of the lighter one, making the binding more stable. Molecules containing bindings of lighter isotopes have a higher tendency to break.
The fractionation factor α, commonly used in geochemistry, is defined via the equilibrium constant K according to Equation (4) (chapter 2.3.2). n is the number of atoms exchanged.

$$\alpha = K^{1/n} \tag{4}$$

Kinetically induced isotope fractionation is based on the mass (m) dependency of reaction speed (v) (Equation 5). As the kinetic energy E_{kin} is usually equal for isotopically heavy and light compounds in the same system, higher masses lead to lower reaction speed and vice versa.

$$E_{kin} = \frac{1}{2} * m * v^2 \tag{5}$$

Therefore isotopic effects are conserved in incomplete unidirectional processes such as evaporation, dissociation, diffusion and biologically induced reactions (Hoefs, 2009).
Using those effects, a variety of methods for several elements was applied to trace sources and processes involving particular elements and molecules,

e.g. hydrogen, oxygen, carbon, nitrogen, sulphur and several heavy elements. An overview is given by Hoefs (2009) as well. With increasing analytical precision a variety of heavy elements can be analyzed on their isotope composition, which widely expands the opportunities of stable isotopes as a geochemical tool. A summary of recent research activities concerning heavy, non-traditional stable isotopes such as Se is given by Wiederhold (2015). This study is concentrating on the Se stable isotope system and its potential as a process tracer in plants.

Selenium has six stable isotopes, ^{74}Se, ^{76}Se, ^{77}Se, ^{78}Se, ^{80}Se and ^{82}Se, with characteristic abundances in the earth crust given in Figure 13. The high range of Se isotopes opens promising possibilities for its application as geochemical tool as well as for precise analytics.

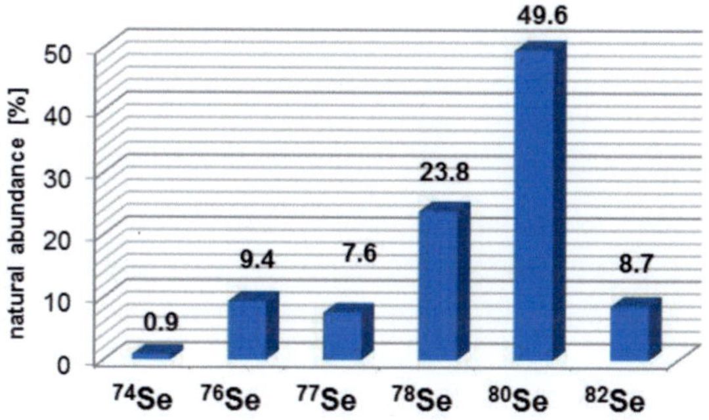

Figure 13: Natural average abundance of the six stable Se isotopes (data from Berglund and Wieser (2011)).

2.3.1 Se stable isotope analytics

Due to a variety of factors Se isotope analytics is quite challenging, e.g. (but not limited to)

- isobaric interferences, especially if argon (Ar) operated machines and hydride generation (HG) techniques are used (Table 2)
- high redox sensitivity of Se, high numbers of redox species with different properties
- high tendencies of Se to volatilize even at low temperatures
- tendency of particular Se species to adsorb to and be retained by various materials, especially plastics
- tendency of Se to adsorb and bind to organic matter
- high ionization potential and therefore low ionization rates, if inductively coupled plasma (ICP) based techniques are used
- generally low concentrations in natural environments

These properties need particular attention regarding analytics, internal correction mechanisms and sample preparation methods (chapters 3.5 and 4). Basic prerequisites for accurate measurements are the reduction of matrix element concentrations and organic residuals to a minimum. Critical elements for the analytics are on the one hand transition metals (e.g. Cr, Co, Cu, Ni, Fe) that inhibit HG (Elwaer and Hintelmann, 2008c), on the other hand hydride generating elements apart from Se (As, Ge) as well as several metal oxides that form isobaric interferences on Se and monitor masses (Fe, Ni, Co, Zn, Cu) (chapter 3.2.2). Main elements (Na, Mg, Al, P, Ca, Fe) may disturb the analytical process because of their quantity and potential interaction with critical elements (chapter 4.4), large organic molecules are in particular harmful for precise analytical measurements within MC-ICP-MS (chapters 4.2 and 4.3). Organic Se molecules are particularly critical, because their behavior – deviating from inorganic Se species (Se oxyanions) – might lead to non-correctable Se isotope fractionation due to selective Se losses within preparation procedures and analytical measurements. Depending on the sample matrix there are several approaches for the digestion and purification of samples (chapters 4.2-4.4). The most suitable preparation methods very much depend on the sample matrix and therefore must be found individually.

A historical overview of analytical techniques applied for Se isotope determinations is given by Johnson (2004). As a first approach, Krouse and Thode (1962) developed a fluorination based technique, SeF_6, in analogy to the applied SF_6 method for sulphur isotope determinations (e.g. Rees, 1978). One disadvantage of this method is the demand of high amounts of Se for reliable measurements, which is generally available for S, but not for Se. The first mass spectrometry based method was published by Wachsmann and Heumann (1992) using thermal ionization mass spectrometry (TIMS). Instrumental mass bias correction using a $^{82}Se/^{74}Se$ Double Spike was first reported by Johnson et al. (1999), whereas the $^{80}Se/^{76}Se$ ratio was given as output. Rouxel et al. (2002) developed a method based on multicollector inductively coupled plasma mass spectrometry (MC-ICP-MS). Using this method, several masses could be measured simultaneously and plasma fluctuations that might have led to invalid and unreproducible results were compensated. On-line HG was applied, which transformed Se from the liquid sample into gaseous H_2Se that was introduced into the plasma. Thereby less energy was needed to ionize the sample, enabling higher plasma temperature and higher ionization rates. Furthermore HG exclusively carries hydride generating elements (Se, As, Ge and some others) into the mass spectrometer, which strengthens the Se signal and avoids matrix interferences. On the other hand, HG might form problematic mass interfering hydrides of Ge and As (Table 2) that require particular correction mechanisms. To correct the instrumental mass bias a sample standard bracketing technique was used, which measures the same standard after and before any sample and subtracts the drift defined by deviation among both standards. This method did not include the correction of matrix effects and artificial isotope fractionation during sample preparation (Rouxel et al., 2002). Zhu et al. (2008b) first used HG-MC-ICP-MS with a $^{74}Se/^{77}Se$ Double Spike, which was able to correct instrumental mass bias as well as fractionation during sample prepa-

ration. This approach combines the advantages of the methods by Johnson et al. (1999) and Rouxel et al. (2002). As all of those methods were quite complex and challenging Layton-Matthews et al. (2006) aimed to set up a method based on classic ICP-MS coupled with a dynamic reaction cell (DRC), HG as sample introduction system (HG-DRC-ICP-MS) and sample-standard bracketing as mass bias correction.
Figures 14 and 15 include the method used for each particular study and data point mentioned. Further variations regarding sample introduction, analytical method and correction mechanisms are given by Layton-Matthews et al. (2006), Elwaer and Hintelmann (2007), Elwaer and Hintelmann (2008a), Elwaer and Hintelmann (2008b), Elwaer and Hintelmann (2008c), Stüeken et al. (2013), Olesik and Gray (2014) and Pogge von Strandmann et al. (2014). Hence, Se isotope determinations are a very challenging and individual issue concerning sample matrix, instrumental setup and required precision. Therefore any laboratory has to find its own method based on the studies mentioned and according to their demands.

2.3.2 Previous studies on Se isotope signatures in the environment

The isotope composition for Se as well as for other isotope systems is commonly given as ratio of two particular isotopes in ‰ related to a standard of a known isotope composition according to Equation (6).

$$\delta^{82}\mathrm{Se}\ [‰] = \left(\frac{(^{82}\mathrm{Se}/^{76}\mathrm{Se})_{\mathrm{sample}}}{(^{82}\mathrm{Se}/^{76}\mathrm{Se})_{\mathrm{standard}}} - 1\right) * 1000 \qquad (6)$$

Isotope fractionation is defined by a particular process and the isotope composition of its states (reagent, product, initial). It can be described by the Rayleigh model that is common in isotope geochemistry (e.g. Mariotti et al., 1981). In an open system with virtually infinite Se pools in reagent as well as product and therefore a linear transport the fractionation can be approximated according to Equation (7). For non-linear transport, the

Rayleigh model requires the knowledge of the amount of Se reacted in the process regarded (1-f), respectively the fraction remaining in the reactant (f) to calculate the fractionation factor (α) and the fractionation Δ (Equations (8) and (9)) (Mariotti et al., 1981).

$$\Delta^{82}\text{Se}\ [‰] = \delta^{82}\text{Se}_{\text{product}} - \delta^{82}\text{Se}_{\text{reactant}} \tag{7}$$

$$\left(\frac{\delta^{82}\text{Se}_{\text{product}}}{\delta^{82}\text{Se}_{\text{initial}}} - 1\right) = f^{(\alpha-1)} \leftrightarrow \alpha = -\frac{(1-f)*(\delta^{82}\text{Se}_{\text{product}}-\delta^{82}\text{Se}_{\text{initial}})}{f*\ln(f)} \tag{8}$$

$$\Delta^{82}\text{Se}\ [‰] = \ln(\alpha) * 1000 \approx (\alpha - 1) * 1000 \tag{9}$$

Varying standards were used depending on the studies and laboratories such as Canyon Diablo Triolite (CDT), MH495 Se standard solution, Merck AAS Se standard solution and NISTSRM3149 (National Institute of Standards and Technology, Gaithersburg – Standard Reference Material 3149) (referred to as NIST3149). To ensure comparability between laboratories, NIST3149 is nowadays the commonly used standard for Se isotope analytics. It is isotopically certified and, as well as the Merck standard, very similar to the average Se isotope composition on earth surface.

Figure 14 shows the Se isotope compositions reported in previously published studies and their authors. Exponents define analytical method (1-6) and standard used (*-****) for each study presented (for comparability all calculated to ^{82}Se/^{76}Se if necessary).

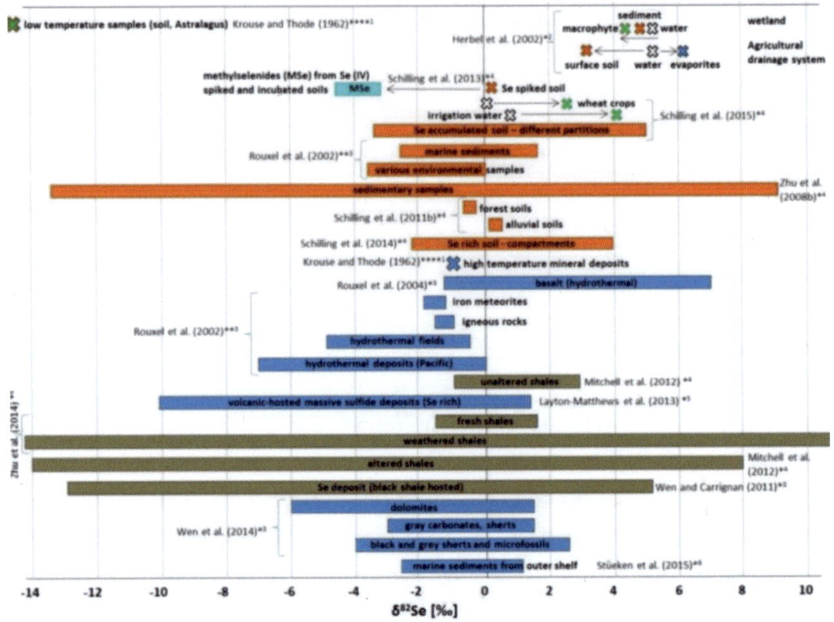

Figure 14: Studies on Se isotope ratios related to a standard material (Equation 6) (* NISTSRM3149 ** MERCK AAS Se standard solution *** MH495 **** Canyon Diablo Troilite meteorite) measured with different methods ([1]SeF6 (Krouse and Thode, 1962); [2]TIMS with 82/74 Double Spike (Johnson et al., 1999); [3]HG-MC-ICP-MS with bracketing (Rouxel et al., 2002); [4]HG-MC-ICP-MS with 74/77 Double Spike (Zhu et al., 2008b); [5]HG-DRC-ICP-MS with bracketing (Layton-Matthews et al. 2006); [6]HG-MC-ICP-MS with bracketing (Stüeken et al. 2013)) and in different sample matrices (cyan – microorganisms, white – water, grey – inorganic laboratory experiments, green – plants, purple – fungi, orange – soils and sediments, dark blue – rock samples, olive – shales) (all Se isotope ratios given in $^{82}Se/^{76}Se$) (studies in which Se isotope fractionation was calculated are excluded from this figure, but presented in Figure 15) (state of November 2015, no claim for completeness).

The Se isotope composition of geological and environmental samples covers a high range from lower than -14 ‰ to more than +11 ‰. Besides rocks originating from high temperature environments, various studies investigated shales and their weathering products regarding the changing Se isotope composition. It is obvious that weathered and altered shales have much higher ranges than the original shales in both directions – enrichment and depletion – in δ^{82}Se. This indicates that the underlying alteration processes are various and frequent as they probably induce a variety of Se fractionations. The same applies for sediments and soils that are exposed to changing environmental conditions. Additionally biosphere compartments such as soil microbes and plants show characteristic Se isotope ratios significantly differing from the surrounding water and soils.

Figure 15 shows Se isotope fractionations for several processes calculated using Rayleigh (Equations (7)-(9)). Exponents were chosen in analogy to Figure 14.

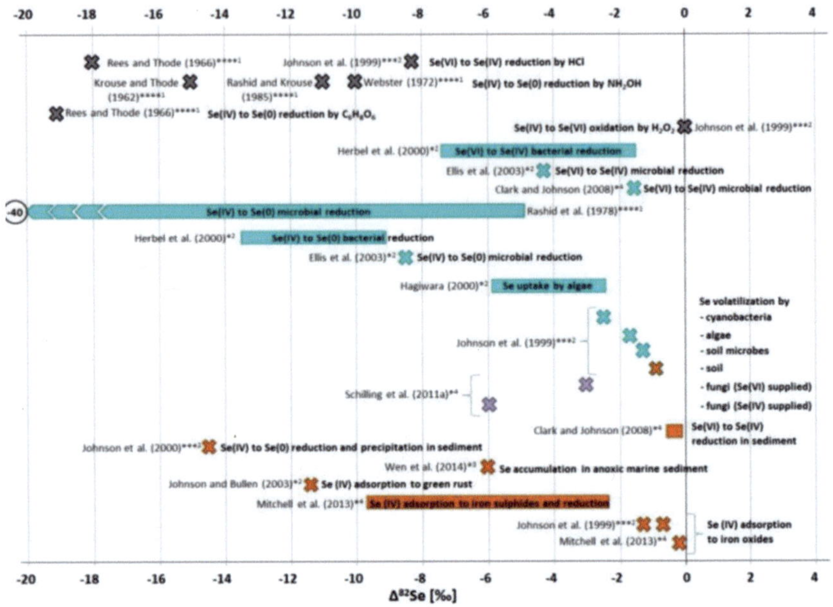

Figure 15: Studies on Se isotope fractionation (calculated using Rayleigh model (Equations (7)-(9)) related to a standard material (* NISTSRM3149 ** MERCK AAS Se standard solution *** MH495 **** Canyon Diablo Troilite meteorite) measured with different methods ([1]SeF6 (Krouse and Thode, 1962); [2]TIMS with $^{82/74}$Se Double Spike (Johnson et al., 1999); [3]HG-MC-ICP-MS with bracketing (Rouxel et al., 2002); [4]HG-MC-ICP-MS with $^{74/77}$Se Double Spike (Zhu et al., 2008b)) and in different sample matrices (cyan – microorganisms, white – water, grey – inorganic laboratory experiments, green – plants, purple – fungi, orange – soils and sediments, dark blue – rock samples, olive – shales) (all Se isotope ratios given in ^{82}Se/^{76}Se) (state of November 2015, no claim for completeness).

A variety of Se isotope fractionating processes in nature have already been figured out. The core process, which is mainly responsible for Se isotope fractionation, is the reduction of Se(VI) (SeO_4^{2-}) to Se(IV) (SeO_3^{2-}) and further to Se(0) and various Se(-II) species. Inorganic reduction combined with sorption or precipitation thereby plays a lesser role than biologically mediated reduction by microorganisms. Reduction of the thermodynamically stable SeO_4^{2-} is energy consuming and therefore particularly leads to isotope fractionation. Transformation of SeO_3^{2-} to Se(0) contains a phase transition towards precipitation of Se(0) or volatilization in addition to reduction itself. In both cases Se(IV) to Se(0) reduction induced higher isotope fractionation than Se(VI) to Se(IV) reduction.

The application of the Se isotope system as a tool to explore geological and environmental processes is on the rise due to analytical progress, increasing awareness regarding the potentials as well as the growing relevance of Se for human health (Figure 16). Therefore the investigation and optimization of analytical and sample preparation processes will probably see an increase in importance as well.

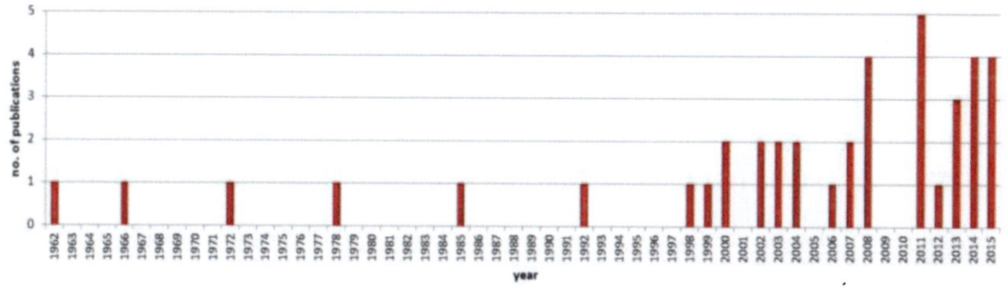

Figure 16: Number of studies exploring or using the Se isotope system in the geological and environmental context, published per year (including analytical studies) (references in Figures 14 and 15 and in chapter 2.3.1) (state of November 2015, no claim for completeness).

Apart from geological and environmental issues there are numerous areas of application for stable Se isotope determinations in various sample matrices and experimental set ups, e.g. for biomedical, forensic and archeological issues, as other stable isotopes are already successfully applied for these purposes (e.g. Heuser and Eisenhauer, 2010; Zangrando et al., 2014).

2.3.3 Se isotope fractionation in plants

As illustrated in Figure 12, biologically mediated reduction plays a major role at Se isotope fractionation in nature. Especially microbial reduction was reported to induce Se isotope fractionation in a wide range, dependent on microbial and Se source species. The difference is most visible with cultivation experiments supplementing Se(VI) versus Se(IV), e.g. with bacterial (Herbel et al., 2000) and microbial consortia (Ellis et al., 2003). The same applies for volatilization processes induced by Se(VI) and Se(IV) supplemented fungi (Schilling et al., 2011b), which showed much higher volatilization rates and higher Se isotope fractionation with Se(IV) (Figure 15). Regarding chapter 2.2, the analogy of those mechanisms to the Se metabolism in plants is obvious and makes Se isotope determinations a promising tool for the investigation of the Se cycle in plants and the soil-plant system. However, there is little data and no systematic studies on Se isotope composition and fractionation in plants. Herbel et al. (2002) determined the Se isotope composition of two wetland plants, the roots of *Scirpus robustus* (saltmarch bulrush) (δ^{82}Se = 3.95 ‰) as well as *Ruppia maritima* (widgeon grass) (δ^{82}Se = 4.73 ‰) and compared it to those of surrounding water and sediment. The average differences to water (-1.11 ‰) and sediment (0.69 ‰) were measurable, but small (Figure 14). According to Herbel et al. (2002) the reduction processes took place deeply within the plants so that the isotopic composition of the total plant could not be influenced significantly. Schilling et al. (2015) determined δ^{82}Se in Se rich wheat crops from Punjab, India, and found a significant enrichment of heavy isotopes within the plants compared to the bioavailable Se fraction in the irrigation water (+2.5 ‰ and +3.2 ‰) (Figure 14). According to the authors this could unlikely be caused by volatilization, instead they assumed alternative processes such as translocations. Both studies show that there are strongly differing mechanisms to induce isotope fractionation. They probably depend on individual ecosystem or plant properties. This indicates that the Se isotope system could be a differentiating tool to examine Se transformation processes in particular environmental systems. However, the number of studies and determined plants is very limited, and in situ environments contain various influence parameters aside from the plant. A differentiation between plant related processes and other Se transformation mechanisms will hardly be possible in a complex in situ system. Controlled cultivation experiments with a limited number of parameters are therefore a reasonable approach to study Se isotope fractionations related to plants depending on particular influence factors. Plant cultivation setups using non-traditional stable isotope systems in order to investigate metabolic processes were already performed for the micronutrients Fe (Guelke and von Blankenburg, 2007; Guelke-Stelling and von Blankenburg, 2012; Arnold et al. 2015), Zn (Arnold et al., 2015) and Cu (Weinstein et al., 2011). Thereby differing uptake and translocation pathways dependent on redox conditions and plant species were revealed and valuable contributions to an adequate Fe, Zn and Cu supply of the food chain were made.

Thus, a variety of factors are indicative for Se isotope applications on plant cultivation setups to be a promising tool for the investigation of the plant's role within the Se cycle as well as the causes for Se accumulations and depletions in plants and the food chain. Those factors are the abundance of Se reduction processes in plants (chapter 2.2), the sensitivity of Se isotope fractionation on reduction in geological, environmental and biological systems as well as the successful application of other non-traditional stable isotope systems in plant cultivation setups (chapter 2.3).

3 Analytical methods

3.1 Overview of standard analytical methods applied

Methods presented in this chapter were applied to determine element concentrations, organic compounds, frequent anions as well as Se species. Standard analytical procedures were available at the Institute of Applied Geosciences (KIT) respectively at University of Basel. Raw data were subtracted on laboratory blanks although this was not reasonable (e.g. to quantify total contamination potentials). Concerned data are marked as such. Averages are given as mean values, standard deviations (±) as 1σ percentile. Calculations were performed using Microsoft Office 2010 Excel.

3.1.1 Element concentrations

For concentration measurements of sodium (Na), magnesium (Mg), aluminium (Al), phosphorus (P), calcium (Ca), chromium (Cr), iron (Fe), cobalt (Co), nickel (Ni), copper (Cu), zinc (Zn), germanium (Ge), arsenic (As) and Se inductively coupled plasma mass spectrometry (ICP-MS) ((X-Series 2, Thermo Scientific) was used. This method is based on the ionization of aqueous atoms and molecules into single positively charged ions by the plasma formed by ionized Ar, a subsequent separation by atomic mass and single detection of the intensity of the ion beam arriving in a Faraday cup. For the determination of Se a collison cell was introduced upstream into the beam. The cell contained a gas mixture from hydrogen (H) and Helium (He), which destroys molecular interferences (ThermoScientific, 2015). This method is very precise and not sensitive on molecular bounds, but affine to isobaric interferences that can be corrected internally via averaging different isotope signals and correct isobaric interferences via the knowledge on natural

isotope abundances. ICP multielement standard VI (Merck) was used as a calibration standard for all elements except P and Ge, which were calibrated with ICP single standards (Merck, Alfa Aeser). In case only Se was measured, Se AAS standard (Roth) was used for calibration.
10 µg L^{-1} of a scandium-rhodium-indium standard solution (*50 µL 1 ppm stock solution to 5 mL sample*) were added to each sample as internal standard. Each measurement series included blanks to control memory effects as well as Se standard solutions and reference materials (CRM-TMDW drinking water reference, High-Purity Standards) for validation and monitoring of instrumental bias. Recoveries of monitoring Se standards (Se = 25 ppb) were on average 100.5 (±1.8) %, those of reference (Se = 10 ppb) 102.5 (±2.6) %.
Samples derived from digestion were diluted 1:10 to 1.3 % HNO_3, the phyto-agar extracts were diluted 1:5 to 0.8 % HNO_3 prior to measurement. Due to complicated matrices, samples derived from the purification experiments were evaporated at 70°C and diluted to 1 % HNO_3.

3.1.2 Total organic carbon (TOC)

The determinations of total organic carbon (TOC) in aqueous samples were performed using a vario TOC cube (elementar). This method is based on a catalytic oxidation of organically bound carbon (C) to up to 1200°C and the measurement of formed CO_2 with non-dispersive infrared sensoring (NDIR). Potassium hydrogen phthalate (Merck) was used as a calibration standard. Standard solutions of 1 ppm were measured continuously for bias monitoring and validation. The average TOC recovery of these standard solutions was 104.0 (±4.2) %. Acidified blanks were measured as well to monitor the influence of acid samples such as digests.
Due to their matrices, all digests, extracts and samples from purification were diluted 1:5 causing an elevated detection limit of 0.9 mg L^{-1}.

3.1.3 Anions

For the determination of frequent anions such as flouride (F^-), chloride (Cl^-), bromide (Br^-), nitrate (NO_3^-), phosphate (PO_4^{3-}) and sulphate (SO_4^{2-}) as well as for qualitative Se species detection in standard solutions, ion chromatography (IC) was used. This method is based on an ion exchange process between mobile phase (eluant) and stationary phase (exchange column). As anions differ in charge and size they are separated chromatically. Anion concentration, type of exchange column, pH value and counter ions in the mobile phase influence the separation process and therefore the suitability for particular anions. An ICS 1000 (Dionex, ThermoScientific) with an IonPac As14 (4 mm) exchange column and an IonPac AG14 pre-column was used. The eluant consisted of 3.5mM Na_2CO_3 and 1.0 mM $NaHCO_3$ with a flow rate of 1.15 mL min^{-1}. The calibration was made from diluted multi ion IC standard solution (Specpure) by Alfa Aeser. Blanks were measured before and after the sequence to trace memory effects.

Although it was not calibrated for the quantification of Se, IC enables the clear differentiation of selenate and selenite peaks in high concentrations when no other anions are present. With this method the selenate and selenite stock solutions were proved to be stable for duration of minimum one year if stored at 4°C. Furthermore Se isotope standards were analyzed on their qualitative species composition.

3.1.4 Se species

Se species were determined in spot tests for selected samples within purification method development and after plant cultivation to monitor the stability of the Se source species. The working group around Markus Lenz from the Institute for Ecopreneurship, located at University of Applied Sciences and Arts Northwestern Switzerland (FHNW), supported this work with their analytics, which are optimized and calibrated for various organic and inorganic Se species. Depending on the sample matrix and the expected species they are using three methods. The first and second one were ion exchange and ion pair, both of which are based on ICP-MS with a subsequent high performance liquid chromatograph (HPLC-ICP-MS) (Bird et al., 1997). Their third method is online pre-concentration, which is based on ICP-MS with a subsequent ion chromatograph (IC-ICP-MS) (Lenz et al., 2012). Ion exchange makes use of the differing sorption potentials of selenate and selenite in a strong anion exchange column, but it is less suitable for organic Se species. Ion pair provides an optimal separation for organic Se species (Bird et al., 1997). Online pre-concentration IC-ICP-MS is a very sensitive and accurate method for the quantification of Se(IV) and Se(VI) in trace concentrations due to an online coupled pre-concentrating trap (Lenz et al., 2012). Sample preparations and data processing were performed according to Bird et al. (1997) and Lenz et al. (2012). A limitation of this method is that samples must be pH neutral. Therefore acid matrices could not be determined on Se species composition without neutralization, which in turn might change Se species stability and composition.

3.2 Development of analytical method for Se isotope determinations

The analytical method for the determination of Se isotopes with HG-MC-ICP-MS was developed and implemented by Ronny Schönberg, Stephan König and their team of the Isotope Geochemistry Group (University of Tübingen). Based on the concept subjected in this study, they set up the analytical foundation as cooperation partners. The development, analytical setting and data assessment are presented in this chapter.

3.2.1 Instrumental setting

Instrumental analytics were based on a Neptune Plus Multicollector ICP-MS (MC-ICP-MS) (ThermoScientific) with on-line HG as sample introduction system. Hydrides were generated from Se (H_2Se) in the liquid samples and introduced into the Ar plasma with a power of 1200 W (chapter 3.2.6). In addition to Ar, methane (CH_4) was injected into the plasma (chapter 3.2.5) (Figure 18).

In analogy to ICP-MS Se hydrides were ionized in the plasma and the ion beam was routed through the electrostatic analyzer (ESA) to focus it and filter out ions that drop out of a given mass to charge ratio. The ion beam was introduced into an electromagnetic field, which separates the ions depending on their masses and introduces them into an array of nine movable Faraday cups. Thereby it was able to measure signal intensities of nine characteristic isotope masses in parallel. Simultaneous detection compensates deviations by plasma fluctuations and therefore highly increases precision. The Faraday cups were connected to electrical ground via adaptable amplifiers that were equipped with high-ohmic resistors of variable resistivity (R). The amplified signal (ion current I) was converted to a tension (V) by a V/F converter according to Equation (10) (Figure 17) (ThermoScientific, 2015). Figure 18 schematically illustrates the analytical procedure as well as the transformation of liquid sample Se into ion beams. The analytical precision for NIST3149 matrix free standard is at least 0.2 ‰.

$$V = I * R \tag{10}$$

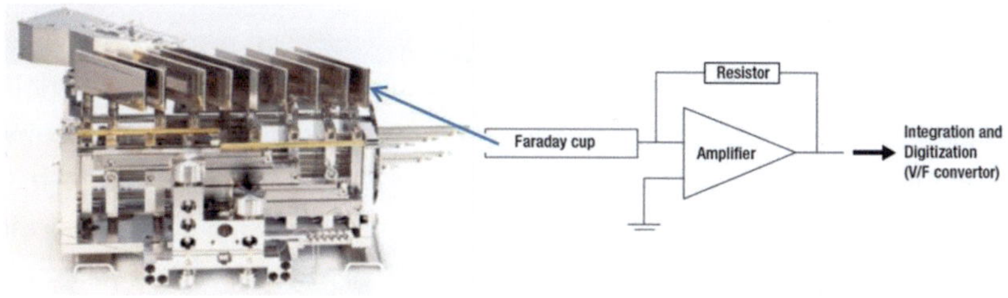

Figure 17: Photograph and scheme of Faraday cups used in MC-ICP-MS analytics (modified from ThermoScientific, 2015).

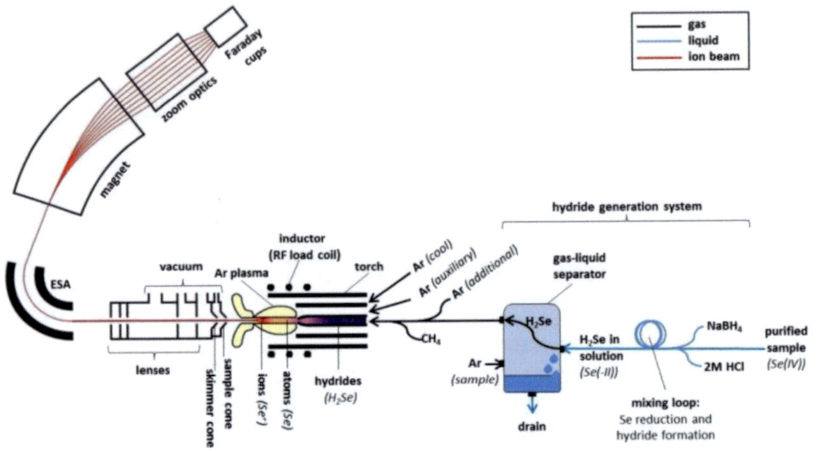

Figure 18: Scheme of HG-MC-ICP-MS analytics for Se isotope measurements (in parts modified from ThermoScientific (2015) and NRM (2015)).

3.2.2 Interferences

As small deviations in isotope composition were supposed to be measured, correction mechanisms as applied for ICP-MS (chapter 3.1.1) were not sufficient. Therefore alternative correction or avoidance mechanisms must be found. Se has a variety of isobaric interferences derived from various sources. Table 2 lists all interferences on Se masses with single isotope abundances of >0.1 % (according to Berglund and Wieser (2011)). It is reasonable to differentiate between those interferences, which are caused by the sample matrix or contaminants from the preparation procedure, which can be filtered out and/or avoided, and those, which are caused by process reagents, e.g. HCl, Ar, Kr, hydrides. The latter are essential for analytics and therefore must be corrected. The first ones are highly critical if they remain residual in the samples, because they do appear neither in the background signal nor in the standard solutions and therefore can directly impact results. Measures to avoid contamination are described in chapter 4.1 and methods to filter out potential interferences in chapter 4.4.

The second ones can be widely corrected by background subtraction (on peak zero). Substantial interferences require additional mathematical correction via simultaneously detected monitor masses, because plasma fluctuations between background and sample measurements might have a significant influence on them. Mathematical corrections using background signal and monitor masses are given in chapter 3.2.8.

Table 2: Interferences on Se and monitor masses measured on Faraday cups at MC-ICP-MS. All potential elemental or molecular interferences with natural isotope abundances of >0.1 % are considered (no regard of ^{17}O, ^{36}S and ^{40}K containing molecules). OPZ – on peak zero (chapter 3.2.8.1), Pur. – purification (chapter 4.4), HG (chapter 3.2.6), calc. – calculations, mathematical correction (chapter 3.2.8).

mass	Target isotope	Interferences							
		from process reagents							
					from sample or contamination				
		ArAr	Kr	ArCl, ClCl	Hyd-rides	Ge	Oxides	other	++ charged
72	^{72}Ge	$^{36}Ar^{36}Ar$		$^{35}Cl^{37}Cl$			$^{56}Fe^{16}O$, $^{54}Fe^{18}O$, $^{40}Ca^{16}O^{16}O$	$^{40}Ar^{32}S$	$^{144}Nd^{++}$ $^{144}Sm^{++}$
73	^{73}Ge			$^{38}Ar^{35}Cl$	^{72}GeH, $^{36}Ar^{36}ArH$		$^{57}Fe^{16}O$, $^{55}Fe^{18}O$, $^{55}Mn^{18}O$	$^{40}Ar^{33}S$	$^{146}Nd^{++}$
74	^{74}Se	$^{38}Ar^{36}Ar$		$^{37}Cl^{37}Cl$	^{73}GeH	^{74}Ge	$^{58}Ni^{16}O$, $^{58}Fe^{16}O$, $^{56}Fe^{18}O$, $^{40}Ca^{16}O^{18}O$, $^{42}Ca^{16}O^{16}O$	$^{40}Ar^{34}S$, $^{39}K^{35}Cl$	$^{148}Sm^{++}$ $^{148}Nd^{++}$
76	^{76}Se	$^{40}Ar^{36}Ar$ $^{38}Ar^{38}Ar$			^{75}AsH	^{76}Ge	$^{60}Ni^{16}O$, $^{58}Ni^{18}O$, $^{44}Ca^{16}O^{16}O$	$^{39}K^{37}Cl$, $^{36}Ar^{40}Ca$, $^{41}K^{35}Cl$	$^{152}Sm^{++}$ $^{152}Gd^{++}$

77	^{77}Se			$^{40}Ar^{37}Cl$	^{76}GeH, $^{40}Ar^{36}ArH$, $^{38}Ar^{38}ArH$, ^{76}SeH		$^{59}Co^{18}O$	$^{38}Ar^{39}K$, $^{36}Ar^{41}K$	$^{154}Gd^{++}$ $^{154}Sm^{++}$
78	^{78}Se	$^{40}Ar^{38}Ar$	^{78}Kr		^{77}SeH		$^{60}Ni^{18}O$, $^{62}Ni^{16}O$	$^{38}Ar^{40}Ca$, $^{41}K^{37}Cl$	$^{156}Dy^{++}$, $^{156}Gd^{++}$
80	$^{40}Ar^{40}Ar$	$^{40}Ar^{40}Ar$	^{80}Kr		^{79}BrH		$^{64}Zn^{16}O$, $^{64}Ni^{16}O$, $^{44}Ca^{18}O^{18}O$, $^{32}S^{16}O^{16}O^{16}O$	$^{40}Ar^{40}Ca$	$^{160}Dy^{++}$ $^{160}Gd^{++}$
81	$^{40}Ar^{40}ArH$				^{80}SeH		$^{65}Cu^{16}O$, $^{63}Cu^{18}O$, $^{33}S^{16}O^{16}O^{16}O$	^{81}Br, $^{36}Ar^{45}Sc$, $^{40}Ar^{41}K$	$^{162}Er^{++}$ $^{162}Dy^{++}$
82	^{82}Se		^{82}Kr		^{81}BrH		$^{66}Zn^{16}O$, $^{64}Zn^{18}O$, $^{32}S^{16}O^{16}O^{18}O$	$^{36}Ar^{46}Ti$, $^{40}Ar^{42}Ca$	$^{164}Dy^{++}$ $^{164}Er^{++}$
83	^{82}SeH		^{82}Kr				$^{67}Zn^{16}O$, $^{65}Cu^{16}O$	$^{38}Ar^{45}Sc$, $^{36}Ar^{47}Ti$, $^{40}Ar^{43}Ca$	$^{166}Er^{++}$
Interference correction		OPZ, calc	OPZ	OPZ	calc	Pur, calc	Pur., HG	Pur, OPZ	Pur, HG

3.2.3 Cup configuration

The cup configuration defines the spatial distribution of the movable Faraday cups and therefore the masses detected. They include Se masses and monitor masses for interference correction. Cup configuration and its intentions are summarized in Table 3. Considering the width of the Faraday cups (1.78 mm), they can gradually be moved to the right position using the Neptune monitoring software. Four out of six Se masses (^{74}Se, ^{77}Se, ^{78}Se and ^{82}Se) were measured for Se isotope ratios in samples and Double Spike correction (chapters 3.5.7 and 3.5.8). ^{76}Se was left out because of its high and superimposing interferences of ^{76}Ge, $^{40}Ar^{36}Ar$ and $^{38}Ar^{38}Ar$ as well as possibly ^{75}AsH. As Ge was one of the most affecting interferences (Table 2),

two masses were monitored to measure not only Ge concentrations, but additionally the instrumental mass bias of Ge for the most valid correction. Mass 80 was monitored mainly for $^{40}Ar^{40}Ar$ to quantify and correct all interferences by all Ar dimers. Mass 81 was measured with the aim of $^{40}Ar^{40}ArH$ quantification to correct interferences by ArAr hydrides. Mass 83 was detected to quantify ^{82}SeH, thereby enabling Se and Ge hydride correction with HG rates of Ge and Se assumed to be equal. Details on mathematical interference correction are given in chapter 3.2.8. Signal amplifiers with a resistivity of 10^{11} Ohm were allocated to all masses except mass 80, which had a lower amplifier of 10^{10} Ohm in order to avoid instrumental damage by the particularly high $^{40}Ar^{40}Ar$ signal, and 83, which was assigned to a stronger one of 10^{12} Ohm because of its low raw signal.
Before starting measurement series, a *gain correction* was conducted, which artificially adds tension on the cups and measures the outcomes to control the signal intensity and - if necessary - includes *gain* factors.

Table 3: Cup configuration applied for Se isotope analytics with HG-MC-ICP-MS

Cup	**L4**	**L3**	**L2**	**L1**	**C**	**H1**	**H2**	**H3**	**H4**
Mass	72	73	74	77	78	80	81	82	83
Target Isotope	^{72}Ge	^{73}Ge	^{74}Se	^{77}Se	^{78}Se	$^{40}Ar^{40}Ar$ ^{80}Se	$^{1}H^{40}Ar^{40}Ar$ $^{1}H^{80}Se$	^{82}Se	$^{1}H^{82}Se$
purpose	Ge correction		Mass bias correction (Double Spike)		smp Se	Ar correction	Ar hydride correction	smp Se	Hydride correction

3.2.4 Signal optimization

To ensure stable and precise mass detection, signals must be tuned in advance. Gas flows (sample Ar, additional Ar) as well as torch and lens positions (Figure 18) were adjusted and optimized before every measurement series. A general enhancement of signal strength and stability was tried by (de-) activating the guard electrode and changing the plasma power (radio frequency (RF)). The first approach led to a significant increase of the signal-to-noise ratio (S/N), which is why the activated guard electrode was maintained. Cold plasma (950 W) resulted in lower S/N as plasma at 1200 W and 1350 W with similar S/N. As plasma fluctuations usually increase with temperature, which in turn is more or less proportional to RF power, 1200 W was chosen for the following measurements.

Testing several cup configurations with pure Se solution, high interferences were observed on masses 75 and 77, which were assumed to be an indication for ArCl ($^{40}Ar^{35}Cl$, $^{40}Ar^{37}Cl$) interferences induced by the use of HCl as reagent. The replacement of 0.2M HCl by 0.2M HNO_3 significantly reduced the signals on Se masses, but did not decrease the interferences respectively the S/N (Appendix IV, Table IV-1). Therefore their origin from ArCl is unlikely. Instead they probably are caused by ArAr hydrides (Table 2), which could later be suppressed by CH_4 and, if necessary, mathematically corrected (chapters 3.2.5 and 3.2.8.1). Signal reduction might have been caused by lower stability of Se(IV) in HNO_3 than in HCl. It was therefore advantageous to retain HCl as process reagent and sample matrix, the Se(IV) stability was at optimum with a molarity of 2M HCl, which was maintained for the further analytical procedure (chapter 3.2.7). All data derived from these tests are listed in Appendix IV (Table IV-1).

3.2.5 Methane injection

As described in chapter 2.3.1, the high ionization potential and therefore low analytical sensitivity of Se as well as various sources of interferences remain challenging. Guo et al. (2013) reported a major extent reduction in the formation of hydrides, Ar dimers, metal oxides and double charged rare earth elements with supplementary injection of methane (CH_4) into the plasma while measuring Se concentrations in biological samples using ICP-MS. Furthermore, the sensitivity concerning Se increased by factor 3.

Tests with CH_4 injection in Se isotope analytics resulted in significant improvements of signal strength, stability and interference reduction, which is why this technique was applied in this study for Se isotope determinations (Figure 18).

3.2.6 Hydride generation

HG was used as sample introduction system instead of liquid injection (Figure 18). This technique has already been applied for Se isotope determinations by Rouxel et al. (2002) and uses the characteristic of Se to form gaseous hydrides in its most reduced form Se(-II) when a sufficient presence of H^+ is given. Se hydrides were generated according to Equation (11) (according to Fitzpatrick et al., 2009):

$$4\,H_2SeO_{3(aq)} + 3\,HCl_{(aq)} + \;3\,NaBH_{4(aq)} \;\rightarrow 4\,H_2Se_{(g)} + 3\,H_3BO_{3(aq)} + 3\,H_2O + 3\,NaCl_{(aq)} \qquad (11)$$

As selenate (Se(VI)) does not take part in the reaction, each sample was pre-reduced to a full transformation of Se into selenite (Se(IV)) by 4M HCl and diluted to 2M HCl afterwards, where selenite is stable in H_2SeO_3 molecules (Olin et al., 2005; Takeno, 2005). Sodium borohydride granulate ($NaBH_4$) was solved in H_2O (2.4 g L^{-1}) and stabilized at pH 11 with 0.1M NaOH, 2M HCl was separately provided. A system of tubes connected to a peristaltic pump continuously merged sample, $NaBH_4$ and HCl to produce Se hydrides according to Equation (11). The tubing systems led the reagents into a loop (to provide sufficient reaction time) and afterwards into a gas-liquid separator (to separate gaseous H_2Se from the residual sample and reaction products) (Figure 18).

More common liquid sample injection systems need additional energy for the evaporation, which lowers the plasma temperature and therefore the ionization rates. As the low ionization rates of Se are a limiting factor, HG keeps the plasma temperature high and thereby enhancing Se ionization rates compared to liquid injection. Additionally residual matrix elements and process reagents are filtered out by the HG process, which decreases the interference potential. Disadvantages such as the remaining transport of other hydride forming elements (As, Ge) and challenges concerning signal stability could be compensated by thorough sample purification, methane injections and accurate Ge correction. That is why on-line HG was chosen as the most suitable sample introduction system.

3.2.7 Double Spike

A variety of Se related processes during sample preparation and analytical measurement might induce artificial Se isotope fractionation in varying extents: (incomplete) ionization, reactive transport within the plasma, sorption, evaporation, incomplete Se recovery etc. In order to analytically separate these isotope fractionations from naturally occurring ones, the Double Spike correction method was applied (Compston and Oversby, 1969; Heuser et al., 2002; Rudge et al., 2009, Johnson et al., 1999; Zhu et al., 2008b and others (Figures 14 and 15)). The Double Spike is a Se standard solution highly enriched in two selected isotopes of low natural abundance, in this case ^{74}Se and ^{77}Se (method after Zhu et al., 2008b). This solution must be added to the sample in a very early preparation stage and fully equilibrated with the sample-Se. Assuming the Double Spike-Se is affected from artificial fractionation processes in an equal extent as the sample-Se, fractionations taking place after Double Spike addition can be approximated and corrected mathematically. Prerequisite is the knowledge of the isotope composition of the Double Spike and its proportion within total-Se (Figure 19) (Compston and Oversby, 1969; Heuser et al., 2002). By having a fixed and known $^{74}Se/^{77}Se$ isotope ratio in the Double Spike solution, the instrumental mass bias can be derived from the sample-Double Spike mixture, and subsequently from the sample. Afterwards this can be used to correct the instrumental output in order to get the original Se isotope composition of the sample, respectively its deviation from a certain standard, defined as $\delta^{82}Se$ (Equation (6)) (Figure 19 - *reduction and inversion*). Based on this data processing (chapter 3.2.8), the instrumental output values were differentiated into an instrumental part ($\beta_{instr,Se}$) induced by analytics and sample preparation and into a natural part ($\beta_{nat,Se}$) induced by the environmental processes aimed at. Figure 19 illustrates the Double Spike method. Further theoretical background is given e.g. by Compston and Oversby (1969) and Rudge et al. (2009).

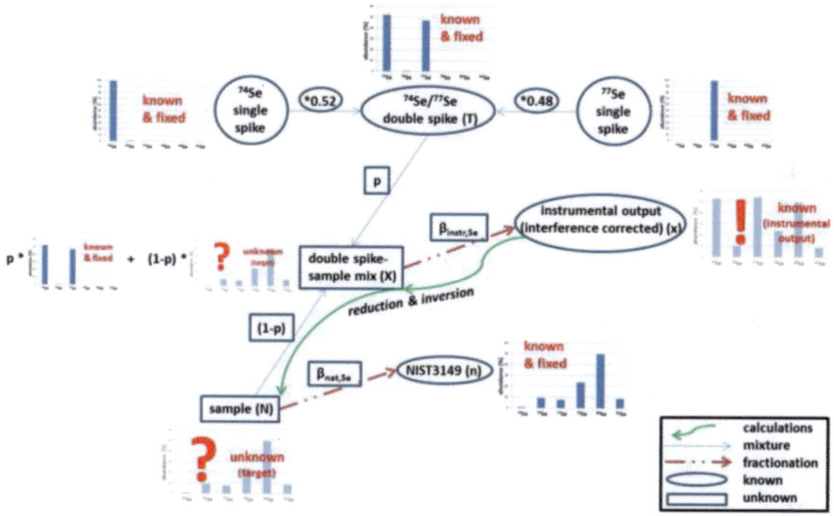

Figure 19: Schematic diagram of the Double Spike technique for the $^{74}Se/^{77}Se$ Double Spike (modified from Rudge et al., 2009).

The inversion calculations of the instrumental output signal (x) requires four Se isotopes in order to gain the original isotope composition of the Double Spike-sample mix (X) and the sample (N) (Figure 19). Based on the remarks in chapter 3.2.3, ^{74}Se, ^{77}Se, ^{78}Se and ^{82}Se were chosen for the inversion. According to Rudge et al. (2009), who set up a MATLAB based model on Double Spikes in several isotope systems, the optimum proportion of $^{74}Se/^{77}Se$ Double Spike with the minimum error probability was 52 % ^{74}Se and 48 % ^{77}Se. A Double Spike-/sample-Se proportion of 1:1 was calculated as the optimum with a relatively wide tolerance range. That is why moderate over- or under-spiking is less critical. Figure 20 shows the dependency of error dimension on the proportion of ^{74}Se and ^{77}Se within the Double Spike as well as the Se proportion of the Double Spike in the sample Double Spike mixture, exemplarily for inversion with ^{74}Se, ^{76}Se, ^{77}Se and ^{82}Se.

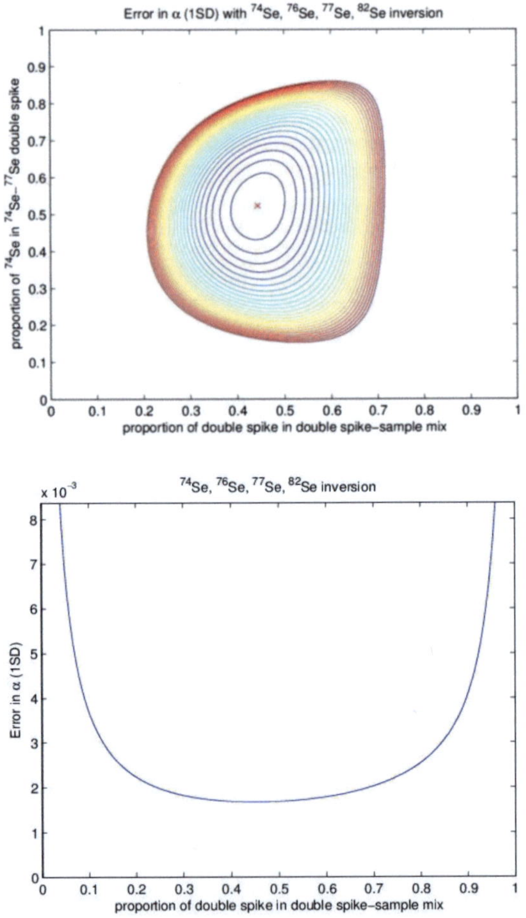

Figure 20: Graph of error dimension depending on the $^{74}Se/^{77}Se$ mix in the Double Spike and the Double Spike/sample mix; calculated with MatLab (provided by John Rudge, personal communication) (exemplarily for inversion with ^{74}Se, ^{76}Se, ^{77}Se and ^{82}Se).

The Double Spike solution was produced from 8.01 mg of enriched (99.1 ±0.2 %) ^{74}Se powder and 17.07 mg of enriched (99.2 %) ^{77}Se (both produced by Chemotrade, Germany) each separately taken up in 2 mL conc. HNO_3 and heated up for 2 h at 95°C on a heating plate to ensure complete solution. Both ^{74}Se and ^{77}Se concentrates were diluted to 0.1M HNO_3 stock solutions with Se concentrations of 16.045 µg g^{-1} (^{74}Se) respectively 33.910 µg g^{-1} (^{77}Se). To create a mixture meeting the requirements calculated by Zhu et al. (2008b) and Rudge (2013) (Figure 20), a Double Spike solution of 52 % ^{74}Se + 48 % ^{77}Se, which equals 324.64 g of ^{74}Se stock solution and 141.35 g of ^{77}Se stock solution, were weighted and mixed. The mixture was equilibrated overnight and diluted to a 10 ppm (10.004 µg g^{-1}) working solution in 0.1M HNO_3 matrix. The isotope composition of the Double Spike solution used in this study is given in Figure 21, its deviation from the optimum distribution was due to slight impurities in the enriched Se isotope powder. Those are included into inversion calculations and therefore do not affect the correction mechanism significantly, as shown for lead (Pb) isotope analytics by Compston and Oversby (1969).

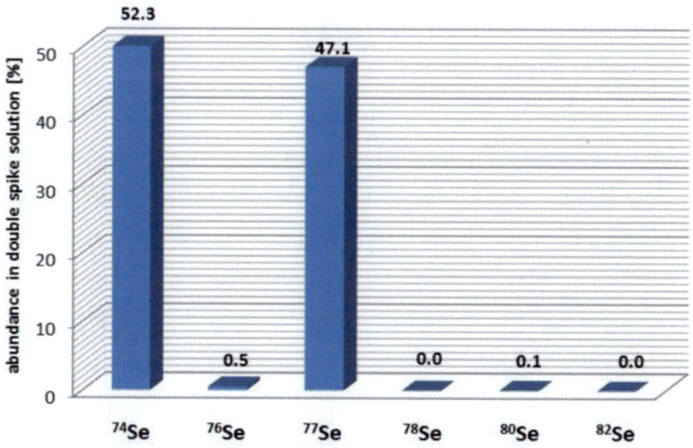

Figure 21: Se isotope composition of the Double Spike solution that is added prior to sample preparation to correct any instrumental mass bias and preparation caused "artificial" Se isotope fractionation (Chemotrade certificate, 2012).

Se species composition of the Double Spike

Selenate and selenite are highly differing regarding their reactivity as well as the rate of oxygen exchange. Selenate is very stable and underlies an exceptionally low oxygen exchange, whereas selenite is very reactive and has high oxygen exchange rates (Kaneko and Poulson, 2012). That is why isotopic equilibration of Double Spike- and sample-Se requires Se to be available in the same species. To monitor Se species prevalence, their availability in the Double Spike was measured with IC. Selenate and selenite were the only species expected in the solution due to its purity, the HNO_3 matrix and oxic conditions (Takeno, 2005). Because the IC column was not suitable for Se oxyanions, the determination was made qualitatively by comparison to high concentrated pure selenate and selenite solutions with additional doping of the Double Spike solution with selenite. To examine the stability of the Double Spike working solutions that were opened several times during experimental phases, two Double Spike solutions – both derived from the stock – were compared: one had been in use for 1.5 years (old), the other was freshly taken from the stock (new). Finding a compromise between resolving the disturbances by high nitrate concentrations due to HNO_3 matrix and low Se concentrations compared to frequent anions, dilution factors 2, 10 and 50 were tested.

The results are given in Figure 22. They show that the very dominant – effectively the exclusive – species was selenite. No selenate peak was detectable, neither in the old nor in the new solution. This indicates the high stability of selenite in this matrix. On the other hand, the selenite peak could clearly be identified with the help of selenite doping and detected in the Double Spike solution as the only anion available despite of nitrate and chloride as process and matrix reagents. A difference between old and new Double Spike solution could not be detected, which indicates a high Se species stability. Based on this knowledge, standard- and sample-Se was fully transformed into selenite before Double Spike adding. Standards and samples were additionally pre-reduced together with the Double Spike as it was part of sample preparation anyway.

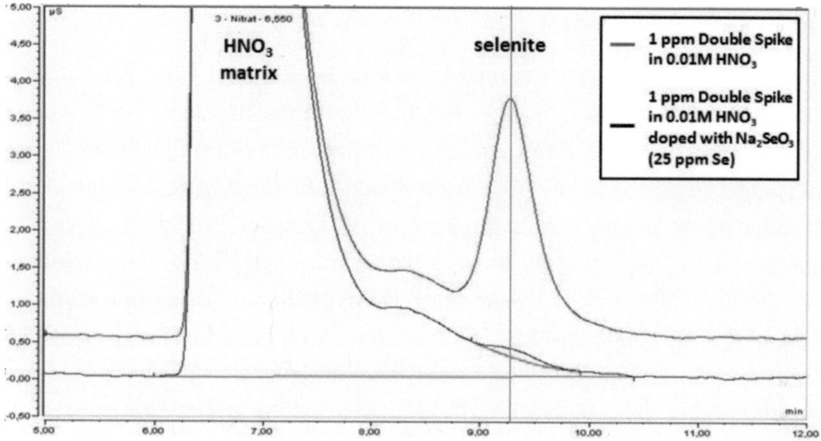

Figure 22: Se species determinations in the Double Spike solution with IC – the highest peak was caused by nitrate from the HNO_3 matrix, the only other peak was proved to be selenite by doping the Double Spike with pure selenite solution in high concentration.

NIST3149

NIST3149 is a standard certified on Se isotope composition commonly used within Se isotope analytics. It was used as the basis for method calibration, validation and $\delta^{82}Se$ calculation of all samples measured. Two diluted NIST3149 solutions, differing in Se species (Se(VI) and Se(IV)) were produced from the NIST3149 stock solution (10,000 µg g^{-1}, 10 % HNO_3, solved from Se(0)). For Se(VI) 100 µg stock solution were diluted to 499.9 g 0.1M HCl. In pure Se solutions with no other counter ions but H^+, weak HCl stabilizes Se(VI), which was assumed to be the dominant species in the stock solution due to its strongly oxidizing HNO_3 matrix, as $HSeO_4^-$ or SeO_4^{2-}. For the NIST3149 in Se(IV) 100 µL stock solution were reduced to Se(IV) by adding 5 mL 6M HCl and heating up at 95°C for 1 h. After cooling down it was diluted to 500 g 2M HCl, which stabilizes Se(IV) as H_2SeO_3(aq) (Dr. Kathrin Schilling, personal comment). The Se concentrations in the two NIST solutions were 2.164 µg g^{-1} (Se(VI)) and 2.196 µg g^{-1} (Se(IV)).

Double Spike calibration

The Double Spike solution can be calibrated by determining several Double Spike-NIST3149 mixtures with varying proportions (Dodson, 1963; Rudge, 2009). Knowledge about the Se isotope composition of Double Spike (T) and NIST3149 (n) as well as the mixing proportion (λ) (Figure 19) enables a calculation of the mixture's Se isotope composition, which then can be compared to the actual Se isotope composition measured and corrected. The Double Spike was well calibrated, if the calculated Se isotope composition of the Double Spike-NIST mixture (Equation (12), left side) equals its measured Se isotope composition (x) corrected by the instrumental mass bias (Equation (12), right side). Calculations can be performed with other isotope pairs as well, if their abundance is significantly above zero (Rudge et al., 2009).

$$\lambda * \mathrm{T} + (1 - \lambda) * \mathrm{n} = \mathrm{x} * \mathrm{e}^{-\beta_{\mathrm{instr,Se}} * \ln\left(\frac{\mathrm{m}(^{74}\mathrm{Se})}{\mathrm{m}(^{77}\mathrm{Se})}\right)} \quad (12)$$

Five solutions with different Double Spike-NIST3149 ratios (λ = 0.5, 0.75, 1, 1.25, 1.5) with Se concentrations of 10 ppb were made. The mixtures were equilibrated and evaporated at 70°C, afterwards taken up in 2M HCl, measured and processed with the inversion calculations described in chapter 3.2.8.2.

3.2.8 Data processing

For the setup of the Se isotope analytics, standards were measured using 90 cycles. Afterwards, all samples, standards and blanks were measured in 40 cycles. Each cycle was corrected and inverted individually and averaged afterwards. The target ratio $\delta^{82/76}$Se was averaged using a 2*SD filter, which excludes all values exceeding the average ± the doubled standard deviation range.

3.2.8.1 Correction of isobaric interferences

Isobaric interferences on Se and monitor masses were corrected in four steps. Firstly, the blank signals, secondly the Ar dimers, thirdly the hydrides and fourthly the Ge induced signals were subtracted from the output (raw) signals. All natural isotope abundances and masses used are provided in Appendix IV, Table IV-2.

On peak zero subtraction

As a first interference correction step the instrumental noise was corrected by subtracting blank signals (*on peak zero*). To include the instrumental fluctuations, blank samples were measured before and after each sample, the average signal of those two blanks was automatically subtracted by the Neptune software (*bracketing*). High background interferences such as ArAr were reduced to a minimum by this; the residuals were corrected as described in chapter 3.2.8.1. Furthermore, trace impurities in Ar (e.g. Kr) and HCl (e.g. Br) were eliminated by on peak zero subtraction. For a valid transferability of blank signals on sample background signals, the HCl matrix had to be of exactly equal molarity. To ensure this, it derived from one batch for all samples, standards and blanks.

ArAr corrections

Ar has three stable isotopes, ^{36}Ar, ^{38}Ar and ^{40}Ar, with varying abundances. Under plasma conditions, Ar tends to form dimers that interfere with several Se and monitor masses, in particular on mass 72 ($^{36}Ar^{36}Ar$), 74 ($^{36}Ar^{38}Ar$), 76 ($^{36}Ar^{40}Ar$ and $^{38}Ar^{38}Ar$), 78 ($^{38}Ar^{40}Ar$) and 80 ($^{40}Ar^{40}Ar$). Any of the five masses must be corrected on ArAr interferences before any further calculations. As $^{40}Ar^{40}Ar$ is the most frequent and therefore most stable one, it is calculated first by using the signal intensity on mass 80 (Equation (14)) corrected by the ^{80}Se fraction, calculated from ^{82}Se (Equation (13)). The determination of the instrumental fractionation factor for Se necessary for the calculation of ^{80}Se, $\beta_{instr,Se}$, is given in Equation (50).

$$^{80}Se(calc) = \left(\frac{NISTab^{80}Se/\ NISTab^{82}Se}{(m(^{80}Se)/m(^{82}Se))^{\beta(instr,Se)}}\right) * I(82) \tag{13}$$

$$^{40}Ar^{40}Ar(calc) = I(80) - \ ^{80}Se(calc) \tag{14}$$

Assuming natural abundances (index *ab*) of Ar isotopes and dimer prevalence, Ar dimers can be calculated on each mass (m) concerned (Equations (15), (17), (19) and (21)) and subsequently correct other masses on Ar dimers (Equations (16), (18), (20), (22), (23)).

<u>$^{36}Ar^{36}Ar$ on I(72):</u>

$$^{36}Ar^{36}Ar(calc) = \ ^{40}Ar^{40}Ar(calc) * \frac{nat.ab^{36}Ar^{36}Ar}{nat.ab^{40}Ar^{40}Ar} \tag{15}$$

$$I(72)corr_{ArAr} = I(72) - \ ^{36}Ar^{36}Ar \tag{16}$$

<u>$^{36}Ar^{38}Ar$ on I(74):</u>

$$^{36}Ar^{38}Ar(calc) = \ ^{40}Ar^{40}Ar(calc) * \frac{nat.ab^{36}Ar^{38}Ar}{nat.ab^{40}Ar^{40}Ar} \tag{17}$$

$$I(74)corr_{ArAr} = I(74) - \ ^{36}Ar^{38}Ar \tag{18}$$

<u>$^{36}Ar^{40}Ar$ and $^{38}Ar^{38}Ar$ on I(76):</u>

$$^{76}(ArAr)(calc) = (^{36}Ar^{40}Ar + \ ^{38}Ar^{38}Ar)(calc) =$$

$$^{40}Ar^{40}Ar(calc) * \frac{(nat.ab^{36}Ar^{40}Ar + nat.ab^{38}Ar^{38}Ar)}{nat.ab^{40}Ar^{40}Ar} \tag{19}$$

$$I(76)corr_{ArAr} = I(76) - (^{36}Ar^{40}Ar + \ ^{38}Ar^{38}Ar)(calc) \tag{20}$$

$^{38}Ar^{40}Ar$ on I(78):

$$^{38}Ar^{40}Ar(calc) = {^{40}Ar^{40}Ar(calc)} * \frac{nat.ab^{38}Ar^{40}Ar}{nat.ab^{40}Ar^{40}Ar} \tag{21}$$

$$I(78)corr_{ArAr} = I(78) - {^{38}Ar^{40}Ar} \tag{22}$$

$^{40}Ar^{40}Ar$ on I(80):

$$I(80)corr_{ArAr} = {^{80}Se(calc)} \text{ (Equation (13))} \tag{23}$$

Hydride correction

Se, Ge and ArAr are the critical compounds, tending to form hydrides that might remain stable in the plasma and therefore act as isobaric interferences (Table 2). Methane injection suppressed hydride formation to a minimum. The residuals were monitored using masses 81 and 83, and hydride correction was optional for high hydride abundances. Hydride formations were assumed not to be dependent on the isotope mass, and rates of Se and Ge were assumed to be comparable.

The formation rate of Se and Ge hydrides (HR) was calculated using I(83) according to Equation (24). The hydride formation rate of ArAr was determined via I(81) according to Equation (25).

$$HR_{SeH/GeH} = \frac{^{82}SeH}{^{82}Se} = \frac{I(83)}{I(82)} \tag{24}$$

$$HR_{ArArH} = \frac{^{40}Ar^{40}ArH}{^{40}Ar^{40}Ar} = \frac{I(81)corr_H}{I(80)corr_{ArAr}} \tag{25}$$

$I(81)corr_H$ is given in Equation (36). The absolute fraction of interfering hydrides in the mass signal could then be calculated using $HR_{SeH/GeH}$ via Equations (26), (28), (30), (31), (32) and (34), the signals were corrected on hydride interferences according to Equations (27), (29), (33), (35) and (37).

<u>^{72}GeH on I(73):</u>

$$^{72}GeH = HR_{\frac{SeH}{GeH}} * I(73) \quad (26)$$

$$I(73)corr_H = I(73) - \ ^{74}GeH \quad (27)$$

<u>^{73}GeH on I(74):</u>

$$^{73}GeH = HR_{SeH/GeH} * I(73)corr_H \quad (28)$$

$$I(74)corr_{ArAr/H} = I(74)corr_{ArAr} - \ ^{73}GeH \quad (29)$$

<u>^{76}SeH, ^{76}GeH, ^{36}Ar^{40}ArH and ^{38}Ar^{38}ArH on I(77):</u>

$$^{76}SeH = HR_{SeH/GeH} * \frac{\left(\frac{NISTab^{76}Se/\ NISTab^{77}Se}{(m(^{76}Se)/m(^{77}Se))^{\beta(Se,nat)}}\right)}{(m(^{76}Se)/m(^{77}Se))^{\beta(Se,instr)}} * I(77)corr_H \quad (30)$$

$$^{76}GeH = HR_{SeH/GeH} * \left(\frac{nat.ab^{76}Ge}{nat.ab^{73}Ge}\right) * I(73)corr_H \quad (31)$$

$$^{77}(ArArH) = \ ^{36}Ar^{40}ArH + \ ^{38}Ar^{38}ArH$$

$$= \frac{I(81)corr_H}{nat.ab^{40}Ar^{40}ArH} * (nat.ab^{36}Ar^{40}ArH + nat.ab^{38}Ar^{38}ArH) \quad (32)$$

$$\boldsymbol{I(77)corr_H = I(77) - \ ^{76}SeH - \ ^{76}GeH - \ ^{77}(ArArH)} \quad (33)$$

<u>^{77}SeH on I(78):</u>

$$^{77}SeH = I(77)corr_H * HR_{SeH/GeH} \quad (34)$$

$$\boldsymbol{I(78)corr_{ArAr/H} = I(78)corr_{ArAr} - \ ^{77}SeH} \quad (35)$$

^{80}SeH on I(81):

$$I(81)corr_H = I(80)corr_{ArAr} * HR_{SeH/GeH} = {}^{80}Se(calc) * HR_{SeH/GeH} \quad (36)$$

^{82}SeH on I(83):

$$I(83)corr_H = I(82) * HR_{SeH/GeH} \quad (37)$$

Germanium correction

As Ge is a very critical element with a variety of potential isobaric interferences, its instrumental fractionation factor was taken into account to increase the precision of interference correction (Equation (38)). For high Ge contents related to Se ($^{73}Ge/^{82}Se > 0.003$) the particular Ge fractionation factor $\beta_{instr,Ge}$ was calculated from I(72) and I(73)corr$_H$, for low Ge contents ($^{73}Ge/^{82}Se < 0.003$) the calculation of $\beta_{instr,Ge}$ was of low reliability and $\beta_{instr,Se}$ was applied instead (Equation (50)). Ge was then corrected using Equations (39) and (40).

$$\beta_{instr,Ge} = \frac{\ln(nat.ab(^{73}Ge/^{72}Ge)/(I(73)corr_H/I(72)corr_{ArAr/H}))}{\ln(m(^{73}Ge)/m(^{72}Ge))} \quad (38)$$

^{74}Ge on ^{74}Se:

$$^{74}Ge = I(73)corr_{ArAr/H} * \frac{nat.ab^{74}Ge/nat.ab^{73}Ge}{(m(^{74}Ge)/m(^{73}Ge))^{\beta(instr,Ge/Se)}} \quad (39)$$

$$\boldsymbol{I(74)corr_{ArAr/H/Ge} = I(74)corr_{ArAr/H} - {}^{74}Ge} \quad (40)$$

The corrected Se masses needed for inversion are given in Equations (33), (35) and (40) (bold equations) as summarized via Equations (41)-(43). In this analytical setup, the signal on mass 82 is effectively interference free after on peak zero correction and can therefore be defined as analogous to ^{82}Se (Equation (44)).

$$^{74}Se = I(74)corr_{ArAr/H/Ge} \quad (41)$$

$$^{77}Se = I(77)corr_{H} \quad (42)$$

$$^{78}Se = I(78)corr_{ArAr/H} \quad (43)$$

$$^{82}Se = I(82) \quad (44)$$

3.2.8.2 Mass bias correction

Inversion calculations were based on iterative circle calculations automatically conducted by MS Office 2010 Excel with a maximum iteration number of 100 and a maximum change of 10^{-9}. Key equations (bold) can only be approximated iteratively, because they directly depend on each other's parameters. All natural isotope abundancies and masses used are provided in Appendix IV, Table IV-2.

All corrected ratios measured ($^{74}Se/^{78}Se$, $^{76}Se/^{78}Se$, $^{77}Se/^{78}Se$, $^{80}Se/^{78}Se$ and $^{82}Se/^{78}Se$) were iteratively reduced to separate the $^{82}Se/^{76}Se$ and $^{82}Se/^{78}Se$ ratios in the sample from the added Double Spike and thereby corrected on mass bias (Zhu et al., 2008b). The parameter names were chosen in analogy to Figure 19:

$\beta_{instr,Se}$	instrumental fractionation factor
$\beta_{nat,Se}$	natural fractionation factor
c [ppm]	concentration
m [amu]	atomic mass
M [g]	Double Spike or sample mass
ab	relative isotope abundance

indices

T	Double Spike
x	Double Spike-sample mix (measured)
X	Double Spike-sample mix (corrected by instrumental fractionation)
n	NIST3149
N	original sample

Equation (45) gives the calculation procedure of the $^{74}Se/^{77}Se$ ratio isotopically fractionated by the instrument including the Double Spike part and the sample part. It makes use of the knowledge on original Double Spike isotope composition and the natural isotope composition, approached by NIST3149 and iteratively approximated to the natural sample isotope composition (Equation (46)). The factor Q74/77 hereby includes fractionations close to natural samples with differing isotope abundances using isotope systems that are frequent in natural samples (Equations (47)-(49)). The natural fractionation factor $\beta_{nat,Se}$ is given in Equation (56).

$$(^{74}Se/^{77}Se)_X = \frac{((^{74}Se/^{77}Se)_T + Q74/77 * (^{74}Se/^{77}Se)_N)}{(1+Q74/77)} \tag{45}$$

$$(^{74}Se/^{77}Se)_N = \frac{(^{74}Se/^{77}Se)_n}{(\frac{m(^{74}Se)}{m(^{77}Se)})^{\beta_{nat,Se}}} \tag{46}$$

$$Q74 = \frac{((^{78}Se/^{77}Se)_X - (^{78}Se/^{77}Se)_T)}{((^{78}Se/^{77}Se)_N - (^{78}Se/^{77}Se)_X)} \tag{47}$$

$$Q77 = \frac{((^{82}Se/^{77}Se)_X - (^{82}Se/^{77}Se)_T)}{((^{82}Se/^{77}Se)_N - (^{82}Se/^{77}Se)_X)} \tag{48}$$

$$Q74/77 = \frac{Q74+Q77}{2} \tag{49}$$

The instrumental fractionation factor $\beta_{instr,Se}$ was calculated using Equation (50), the instrumental fractionation of all isotope ratios in the mixture can then be calculated using this factor (Equations (51)-(53)).

$$\beta_{instr,Se} = \frac{\ln(\frac{(^{74}Se/^{77}Se)_X}{(^{74}Se/^{77}Se)_x})}{\ln(\frac{m(^{74}Se)}{m(^{77}Se)})} \tag{50}$$

$$(^{78}Se/^{77}Se)_X = (^{78}Se/^{77}Se)_x * (\frac{m(^{78}Se)}{m(^{77}Se)})^{\beta_{instr,Se}} \tag{51}$$

$$({}^{82}Se/{}^{77}Se)_X = ({}^{82}Se/{}^{77}Se)_x * (\frac{m({}^{82}Se)}{m({}^{77}Se)})^{\beta_{instr,Se}} \quad (52)$$

$$({}^{82}Se/{}^{78}Se)_X = \frac{({}^{82}Se/{}^{77}Se)_X}{({}^{78}Se/{}^{77}Se)_X} = ({}^{82}Se/{}^{78}Se)_x * (\frac{m({}^{82}Se)}{m({}^{78}Se)})^{\beta_{instr,Se}} \quad (53)$$

To calculate the isotope composition of the sample (without Double Spike fraction) (Equation (54)), $^{82}Se/^{78}Se$ was used. This isotope pair was hardly available in the Double Spike, but occurs predominantly in natural samples. It was necessary to include another factor (Q78) into the calculations to take particular abundances in the Double Spike – relative to natural ones – into consideration (Equation (55)).

$$({}^{82}Se/{}^{78}Se)_N = ({}^{82}Se/{}^{78}Se)_X + \frac{({}^{82}Se/{}^{78}Se)_X - ({}^{82}Se/{}^{78}Se)_T}{Q78} \quad (54)$$

$$Q78 = Q74/77 * \frac{({}^{78}Se/{}^{77}Se)_N}{({}^{78}Se/{}^{77}Se)_T} \quad (55)$$

The natural fractionation factor (Equation (56)) can be derived from the fractionation of $^{82}Se/^{78}Se$ in the sample (Equation (54)). The natural isotope abundance was hereby approached by NIST3149 and iteratively approximated to the original sample isotope ratios. The isotope ratios of the sample can thereby be calculated (Equations (57)-(63)).

$$\boldsymbol{\beta}_{nat,Se} = \frac{\ln(\frac{({}^{82}Se/{}^{78}Se)_n}{({}^{82}Se/{}^{78}Se)_N})}{\ln(\frac{m({}^{82}Se)}{m({}^{78}Se)})} \quad (56)$$

$$({}^{74}Se/{}^{77}Se)_N = \frac{({}^{74}Se/{}^{77}Se)_n}{(\frac{m({}^{74}Se)}{m({}^{77}Se)})^{\beta_{nat,Se}}} \quad (57)$$

$$({}^{74}Se/{}^{78}Se)_N = \frac{({}^{74}Se/{}^{78}Se)_n}{(\frac{m({}^{74}Se)}{m({}^{78}Se)})^{\beta_{nat,Se}}} \quad (58)$$

$$({}^{76}Se/{}^{77}Se)_N = \frac{({}^{76}Se/{}^{77}Se)_n}{(\frac{m({}^{76}Se)}{m({}^{77}Se)})^{\beta_{nat,Se}}} \tag{59}$$

$$({}^{78}Se/{}^{77}Se)_N = \frac{({}^{78}Se/{}^{77}Se)_n}{(\frac{m({}^{78}Se)}{m({}^{77}Se)})^{\beta_{nat,Se}}} \tag{60}$$

$$({}^{80}Se/{}^{77}Se)_N = \frac{({}^{80}Se/{}^{77}Se)_n}{(\frac{m({}^{80}Se)}{m({}^{77}Se)})^{\beta_{nat,Se}}} \tag{61}$$

$$({}^{82}Se/{}^{77}Se)_N = \frac{({}^{82}Se/{}^{77}Se)_n}{(\frac{m({}^{82}Se)}{m({}^{77}Se)})^{\beta_{nat,Se}}} \tag{62}$$

$$({}^{82}Se/{}^{76}Se)_N = \frac{({}^{82}Se/{}^{76}Se)_n}{(\frac{m({}^{82}Se)}{m({}^{76}Se)})^{\beta_{nat,Se}}} \tag{63}$$

From the calculated original isotope ratios in the sample N, the relative abundance of ^{77}Se was calculated using all ratios (Equation (64)). All other abundances could be derived from there (Equations (65)-(69)).

$$ab^{77}Se_N = \frac{1}{(({}^{74}Se/{}^{77}Se)_N+({}^{76}Se/{}^{77}Se)_N+({}^{78}Se/{}^{77}Se)_N+({}^{80}Se/{}^{77}Se)_N+({}^{82}Se/{}^{77}Se)_N+1)} \tag{64}$$

$$ab^{74}Se_N = ({}^{74}Se/{}^{77}Se)_N * ab^{77}Se_N \tag{65}$$

$$ab^{76}Se_N = ({}^{76}Se/{}^{77}Se)_N * ab^{77}Se_N \tag{66}$$

$$ab^{78}Se_N = ({}^{78}Se/{}^{77}Se)_N * ab^{77}Se_N \tag{67}$$

$$ab^{80}Se_N = ({}^{80}Se/{}^{77}Se)_N * ab^{77}Se_N \tag{68}$$

$$ab^{82}Se_N = ({}^{82}Se/{}^{77}Se)_N * ab^{77}Se_N \tag{69}$$

The isotope ratios could also be described by the commonly used δ, which is defined as the deviation of the sample from NIST3149 concerning a particular isotope ratio, expressed in ‰ (Equation (6)). Equations (70)-(73) give the calculations of δ for each isotope ratio. The $\delta^{82}Se/^{76}Se$ ratio (bold) is the one most commonly published in literature and used in this study (Equation (73)). It is analogous to Equation (6).

$$\delta^{74}Se/^{77}Se\ [‰] = (\frac{(^{74}Se/^{77}Se)_N}{(^{74}Se/^{77}Se)_n} - 1) * 1000 \tag{70}$$

$$\delta^{78}Se/^{77}Se\ [‰] = (\frac{(^{78}Se/^{77}Se)_N}{(^{78}Se/^{77}Se)_n} - 1) * 1000 \tag{71}$$

$$\delta^{82}Se/^{77}Se\ [‰] = (\frac{(^{82}Se/^{77}Se)_N}{(^{82}Se/^{77}Se)_n} - 1) * 1000 \tag{72}$$

$$\boldsymbol{\delta^{82}Se/^{76}Se}\ [‰] = \boldsymbol{\delta^{82}Se}[‰] = (\frac{(\boldsymbol{^{82}Se/^{76}Se})_N}{(\boldsymbol{^{82}Se/^{76}Se})_n} - \mathbf{1}) * \mathbf{1000} \tag{73}$$

Equation (74) gives the atomic weight of Se in the sample calculated via the isotope masses and abundances.

$$m(Se)_N = m(^{74}Se) * ab^{74}Se_N + m(^{76}Se) * ab^{76}Se_N + m(^{77}Se) * ab^{77}Se_N + m(^{78}Se) * ab^{78}Se_N + m(^{80}Se) * ab^{80}Se_N + m(^{82}Se) * ab^{82}Se_N \tag{74}$$

The Se concentration of the sample can be calculated according to Equation (75). M(T) and M(N) are the weights of Double Spike and sample mixed at the beginning.

$$c(Se)_N[ppm] = \frac{(c(Se)_T * M(T) * ab^{78}Se_T) * \frac{\frac{m(Se)_N}{m(Se)_T} * Q78}{ab^{78}Se_N}}{M(N)} \tag{75}$$

4 Producing precise and valid Se isotope data by developing individual sample treatment methods

The application of Se isotope analytics (chapter 3.2) to plant and phytoagar samples that were derived from *Minimum Parameter* cultivation setups (chapter 5) requires previous sample treatment. No methods regarding the demands of those matrices with reference to Se isotope analytics were published yet. That is why applied methods had to be adapted and new ones had to be developed in order to meet the challenges occurring with organic rich samples to gain precise and valid Se isotope data. Subject of chapter 4 is the evaluation of existing and new methods and, as a consequence, the development of a comprehensive procedure for both sample types.

4.1 Purity and cleaning procedures

Trace element analytics in general and Se isotope analytics in particular require high purity of reagents, equipment and workspace to avoid Se as well as other critical contamination. As described in chapters 2.1.1 and 2.3.1.1, Se has adhesive properties and therefore tends to stick onto surfaces in varying extents, dependent on Se species and surface material. Particularly if Se amounts are at trace level, low contaminations can have high impacts on the reliability of the results. Especially the use of the Double Spike might lead to isotopic shifts even at low Se contamination level. This makes the avoidance of Se blanks and therefore sufficient cleaning procedures essential. Furthermore the applicability of an open laboratory for sample preparation was tested. If no clean laboratory was needed, the

entire analytical process was simplified and therefore of improved applicability for this and further studies.
Within this study there was an exclusive application of suprapure grade acids. For the required solid chemicals, only the purest grade available was taken. In-house produced chemicals or those with lack of certified trace element contents were digested and tested on main and critical elements using ICP-MS. A list of all chemicals and reagents used is provided in Appendix I.
The choice of the material for liquid handling and sample storage is an important factor to ensure a working environment with very little contamination. Thorough and individual cleaning procedures that remove contaminants, while not influencing surface properties, are additionally mandatory to retain high quality as well as a clean workspace. Perfluoroalkoxy alkane polymer (PFA) was the preferred material for acid and sample containers, because it has a repelling surface that ensures full sample recovery and no residuals. Furthermore it is both acid and heat resistant (>320°C) enabling cleaning with boiling acid (AHF, 2014). If the use of PFA material was not possible or applicable, glass (borosilicate or DURAN) was preferred, because it has a low adhesive affinity to Se and it is also resistant to heat and acid. Concentrated and double-distilled HNO_3 was used for all cleaning procedures, because – besides of general contaminant solution – it tends to oxidize Se and therefore easily mobilizes it from the surfaces. All PFA beakers and quartz vessels were rinsed with H_2O (Millipore), flushed with Mucasol (1 %), wiped out and rinsed again. Then they were stored in 5 % HNO_3 in a PFA flange container and put on a heating plate (200°C) for five days. After cooling down the HNO_3 was removed, each of the beakers was rinsed with H_2O and stored in H_2O for two more days. Afterwards they were dried at flying air. Scaling beakers, vacuum flasks and ceramic Büchner funnels used for cultivation experiments and phytoagar extraction were thoroughly rinsed with H_2O, 5 % HNO_3 and H_2O again. Each Se species and concentration had its own equipment to minimize contamination.
One-way polyethylene (PE) containers (Patho beakers, centrifuge tubes) were rinsed with H_2O and dried before usage. Contamination from produc-

tion and storage (e.g. dust) could be removed by this. Se plays a minor role in this context. Multiway PE, polypropylene (PP) or polystyrene (PS) equipment (magenta boxes, minicolumns, tweezers) tended to retain Se if exclusively rinsed with H_2O and did not tolerate HNO_3 boiling. Magenta boxes and couplers were particularly affected by Se traces due to Se volatilization during plant cultivation and the affinity of organic volatile Se species to stick to plastic surfaces. They were rinsed several times with H_2O and stored in 5 % HNO_3 for five days at room temperature. Afterwards they were thoroughly rinsed with H_2O again and dried at flying air.

Minicolumns were rinsed with H_2O before first usage. After usage the packing material was fully removed. Thiol cellulose powder (TCP) could easily be flushed out with H_2O, whereas AG1-X8 resin was dried at flying air for several days and then could simply be shaken out afterwards (packing materials described in chapter 4.4.1). The PE frits were blown out of the columns by compressed filtered air, and columns, frits and the adjacent caps as well as tips (if used) were rinsed separatedly with H_2O several times. They were stored in 10 % of an ethanol-isopropanole mixture and mechanically shaken for 30 min. Afterwards they were rinsed with H_2O and stored in 5 % HNO_3 for five days, afterwards stored in H_2O for another two days, rinsed again and dried at flying air. Detailed cleaning instructions for all materials used are provided in Appendix II.

The entire sample preparation was performed in a laboratory hood, which was exclusively used for sample preparation and evaporation. It was cleaned thoroughly with detergents, 5 % HNO_3 and H_2O before and after every experimental series. A metal free heating plate (AHF) and particular equipment was used. It remained within the hood during the experiments and cleaned according to chapter 4.1. Process steps outside the hood (digestion, shaking, centrifugation, fridge storage) were exclusively done in closed containers.

To monitor the contamination potential of the work space, blanks were taken on a regular basis. For this purpose 1 % HNO_3 in an open beaker was placed into the hood during preparation and evaporation phase (5-6 days). Afterwards it was measured for critical elements with ICP-MS. Table 4 shows the hood blanks at three dates within the project timeframe compared to analytical blanks used in ICP-MS measurements in the ISO 1000 clean laboratory at AGW.

Table 4: Workspace blanks – 1 % HNO_3 in hood for 5-6 days (separation procedure + evaporation of all samples) at three times during the experimental phase compared to analytical blanks (1 % HNO_3) used in the ICP-MS clean laboratory at AGW/KIT (ISO 1000), both measured with ICP-MS

Element concentrations [µg L^{-1}]	**ISO 1000 clean lab blank** (n=10)	**workspace blanks (n=1)**		
		11/2012	**10/2013**	**07/2014**
Na	2.24 ±0.81	25.8	37.3	20.1
Mg	0.11 ±0.04	14.8	23.4	11.5
Al	0.16 ±0.07	25.5	20.4	20.9
Ca	2.20 ±0.86	89.5	261	106
Cr	0.00 ±0.09	2.95	0.19	0.05
Fe	0.09 ±0.00	46.5	17.3	15.3
Co	0.00 ±0.03	0.02	0.02	0.00
Ni	0.02 ±0.00	0.47	0.30	0.00
Cu	0.03 ±0.01	1.44	2.03	0.80
Zn	0.36 ±0.01	3.65	8.42	3.90
Ge	0.00 ±0.01	<0.003	<0.003	<0.003
As	0.00 ±0.00	0.08	<0.01	<0.007
Se	0.04 ±0.00	<0.08	<0.06	<0.06

As expected, the air in the exhauster within the open laboratory contains ubiquitary elements in measurable amounts, e.g. Ca, Fe, Zn. Regarding the accumulation time of 5-6 days, concentrations are low in comparison to the process blanks listed in Table 5. Exceptionally critical elements such as Ge, As and Se are in the range or marginally above the analytical blank used in the clean lab.

Table 5 shows process blanks from the three purification procedures described in chapter 4.4. The blanks were processed through all purification steps, respectively, and measured on main and critical elements using ICP-MS afterwards. However, the blanks are an addition of reagent blank, equipment blank and workspace blank (chapter 4.1).

Table 5: Process blanks derived from purification methods (A), (B) and (C) measured with ICP-MS (raw data available in Appendix IV, Tables IV-6 to IV-8)

Element concentrations [µg L^{-1}]	**Method (A) (anion exchange) (n=6)**	**Method (B) (thiol retention) (n=3)**	**Method (C) (hydride separation + anion exchange) (n=2)**
Na	48.5 ±46.1	90.8 ±40.4	1.57 ±0.13
Mg	5.65 ±3.34	23.9 ±0.45	3.39 ±0.76
Al	9.41 ±5.44	21.3 ±6.11	3.18 ±0.18
Ca	50.7 ±27.5	918 ±492	31.1 ±9.81
Cr	0.45 ±0.12	0.32 ±0.02	8.80 ±0.31
Fe	12.1 ±8.07	12.6 ±4.68	1.77 ±0.34
Co	0.03 ±0.01	0.01 ±0.00	0.06 ±0.01
Ni	0.48 ±0.43	0.35 ±0.05	38.2 ±5.14
Cu	0.37 ±0.18	0.80 ±0.04	0.24 ±0.00
Zn	6.24 ±2.93	58.5 ±28.1	1.94 ±0.80
Ge	0.01 ±0.01	0.11 ±0.03	<0.003
As	0.11 ±0.08	0.36 ±0.16	0.05 ±0.00
Se	0.16 ±0.04	0.46 ±0.31	<0.1

The comparably low blank level is a result of the thorough cleaning procedures described in this chapter. Their necessity became obvious due to exceptionally high Se process blanks of up to 10 % of sample Se content (25-38 µg L^{-1}) observed in reused columns within method (B) purification (Appendix IV, Table IV-7). Minicolumns were washed and rinsed thoroughly, but only H_2O based and without frit removal. Se was probably adsorbed to

the column surfaces and retained in frit pores and interstices. They could be removed by treatment according to chapter 4.1. At first, magenta boxes were just thoroughly rinsed with H_2O after usage. After a first reuse the accumulation of Se in phytoagar and plants of the Se free box was observed. About 5 % of the lowest concentration added was determined in plants and phytoagar samples (Appendix IV, Table IV-15). As no Se was added to this box, it probably originated from residuals adsorbed to the surfaces particularly former volatile phases. Subsequently, pure H_2O did not desorb them while extended exposure to HNO_3 oxidized and remobilized the Se. This problem was not experienced again afterwards.
Open laboratories are applicable for sample preparation for Se isotope analytics with some limitations. Ubiquitary element concentrations are elevated compared to clean laboratories, which especially applies for critical metals such as Fe and Zn. They must be kept within acceptable ranges by keeping adequate cleaning instructions, using suprapure reagents as well as monitoring hood and process blanks. For blank tolerance checks concerning the analytical method see chapter 4.5.1. The more important problem is the potentially high Se blank due to its adhesive properties and high impacts on Se isotope ratios by the Double Spike usage. Through adequate choices of material and cleaning procedures the Se blanks can be kept at a low level. HNO_3 is the most suitable cleaning reagent, because it oxidizes and mobilizes Se, therefore being able to desorb Se residuals from surfaces. Additionally it is suitable to solve and remove most critical metals to a high extent.

4.2 Phytoagar treatment

Phytoagar is a semi-solid growth medium of gelatinous consistency frequently used in biological cultivation experiments and applied within the *Minimum Parameter* approach (chapter 5). It is free of nutrients and trace elements, consisting only of the organic molecules agarose (70 %) and agaropectine (30 %) (Duchefa, 2011) as well as the Se added. Raw phytoagar is a powder that dissolves in water at temperatures between 95 and 100°C, forming a liquid solution. If this solution is cooled down to 30-40°C, it turns into a gelatinous and stable semi-solid mass. Phytoagar solution is pH neutral (pH 6.8-7). In this study, 0.4 % phytoagar was used. Its density at room temperature is very close to water (1.010 ±0.003 g cm^{-3} (n=3)).

Phytoagar consists of a solid (phytoagar powder) and a liquid part (H_2O) connected to each other. The solid part forms a lattice in whose interspaces the H_2O molecules are integrated (Figure 23). This composition leads to an effectively solid growth medium in which the water molecules and the solved Se can move freely. According to Davies et al. (2010) there is no sorption to the solid part of the phytoagar, and diffusion within pure water and 0.4 % phytoagar is very similar – the diffusion coefficient in pure water is $2.27*10^{-9}$ m^2s^{-1} compared to $2.25*10^{-9}$ m^2s^{-1} in 0.4 % phytoagar (extrapolated from Davies et al. 2010), making a difference of 0.71 %.

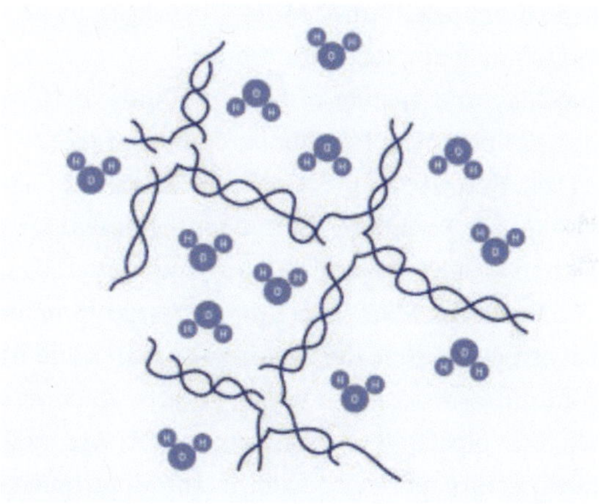

Figure 23: Scheme of organic lattice and connected H_2O molecules of semi-solid phytoagar (Duchefa, 2011).

The analytical methods for the determination of Se and matrix element concentrations, TOC, Se species and Se isotope composition require the availability of samples in aqueous form (chapter 3). To transform semi-solid

phytoagar into a liquid consistency, digestion after Kopp (1999) in analogy to chapter 4.3 and extraction of the liquid phase with vacuum filtration were compared. As sorption of Se to the solid lattice within the phytoagar was negligible (chapter 4.2) (Davies et al., 2010), it was assumed that the entire Se is related to the liquid phase. A vacuum filtration method was developed in order to filter out the Se containing liquid phase without changing the Se concentration, species or isotope composition. The setup contained a vacuum pump (1400 RPM, KNF) connected to a 100 mL filter flask associated with a 120 mL Büchner funnel, and 70 mm diameter membranes with 0.45 µm pore width (Roth) (Figure 24). Therefore, the vacuum in the flask pulls out the liquid fraction from the phytoagar in the funnel, taken from the semi-solid stock and separated by filter. A full separation of liquid and solid phase was not possible. Because of this each sample was treated for a defined time period (15 min) and the extracted sample volume was tested on its suitability to represent the entire phytoagar in Se concentration, species distribution and isotope composition.
To test reproducibility and validity of both methods tests were performed that included the addition of Se into hot liquid phytoagar (70°C) with various concentrations (100, 500, 1000 µg L^{-1}) and species (SeO_4^{2-}, SeO_3^{2-}, org. Se) in analogy to the *Minimum Parameter Experiments* (MinPaX) (chapter 5.1). For digestion after Kopp (1999), 2 mL of the hot liquid phytoagar was pipetted and stored in 7.5M HNO_3, which was later digested in analogy to chapter 4.3. Because it was proved that the densities of water and liquid phytoagar are similar, significant errors due to wrong pipette volumes were improbable. In contrast, the pipetting of solid phytoagar was highly inaccurate; therefore elevated errors were expected in the determination of Se after cultivation compared to the samples taken before. For vacuum filtration, the phytoagar was cooled down to room temperature after Se addition and treated afterwards as described above. As this method was designed for phytoagar at room temperature, no significant differences were expected in treating phytoagar after cultivation. After treatment, digests and extracts were examined for on Se concentration and Se recovery was calculated related to the Se previously added. To avoid contamination, individual flasks

and funnels were used for each concentration and species, and cleaned with 5 % HNO_3 followed by H_2O (millipore) after and before any usage.

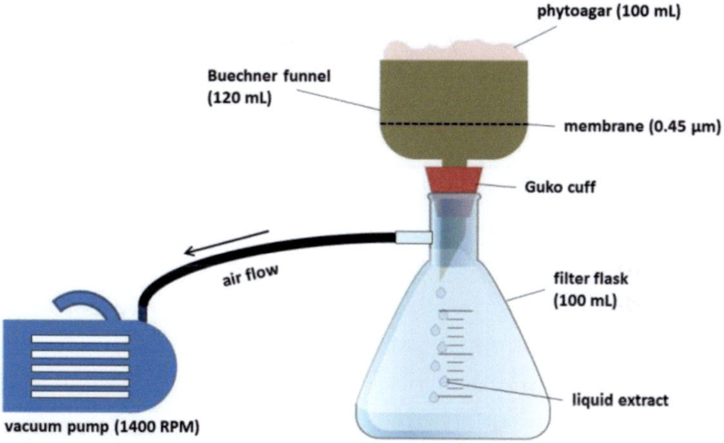

Figure 24: Simplified scheme of the experimental setup for vacuum filtration (not to scale).

For organic destruction, 3 mL conc. HNO_3 and H_2O_2 each were added the liquid extracts and heated up in closed beakers at 80°C for 24 h. Afterwards the mixture was evaporated at 70°C to approx. 500 µL, 3 mL HNO_3 and H_2O_2 were added again and it was evaporated at 70°C until dryness was reached. Figure 25 shows the Se recoveries derived from phytoagar doped with different Se concentrations and species in digestion after Kopp (1999) and vacuum filtration experiments. Table 6 lists the average Se recovery depending on initial Se concentration and species.

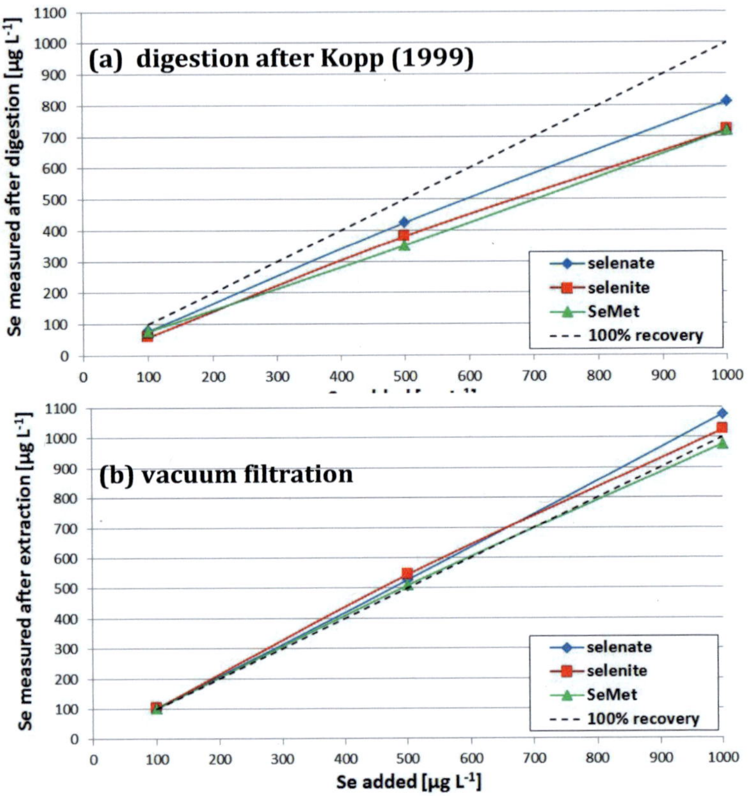

Figure 25: Se recoveries dependent on Se concentration and species added using digestion after Kopp (1999) (a) and vacuum filtration (b).

Digestion resulted in badly reproducible recoveries with deficits of an average 25.7 (±5.5) % and maxima of 42 %, probably due to evaporation and volatilization of Se during digestion. This does not only lead to inaccurate concentration determinations in phytoagar: According to Cappa et al. (2003) the isotope fractionation of hydrogen and oxygen is very high at evaporation processes. Therefore it probably has effects on Se isotope composition, especially with rates in the range measured. There is a slight species de-

pendence regarding selenate yielding higher recoveries than selenite and SeMet, probably because it is a thermodynamically stable molecule that does hardly transform into volatile species (Olin et al., 2005). Any significant dependence on concentration could not be detected.
Vacuum filtration resulted in full Se recoveries (103 (±2.6) %) averaged on concentration and species. Slight species dependent differences can be explained with small variations in solute transport behavior. The diffusion coefficient (Equation (76)), which determines the solute transport speed in solution via 1st Fick's Law (Equation (77)), negatively correlates with the molecular size (which is proportional to the hydrodynamic radius R_0), while the other parameters are species independent (Cussler, 1997). That is the reason why relative recoveries of more than 100 % were detected for the small oxyanions and values around 100 % for the larger organic molecule SeMet.

$$D = \frac{k_B * T}{6 * \pi * \eta * \boldsymbol{R_0}} \tag{76}$$

$$J = -D \frac{\partial c}{\partial x} \tag{77}$$

D [m^2 s^{-1}]	diffusion coefficient
k_B [J K^{-1}]	Boltzmann constant
T [K]	temperature
η [N·s·m^{-2}]	solvent's dynamic viscosity
R_0 [m]	**hydrodynamic radius of the solute**
J [mol m^{-2} s^{-1}]	solute diffusion stream
c [mol m^{-3}]	solute concentration
x	direction

No dependence on concentration was detected. In any case, Se recoveries were all close to 100 % (96-109 %) with low scattering and very good reproducibility (Table 6). Therefore vacuum filtration can be characterized as a precise method to determine Se concentrations in phytoagar, to monitor Se species stability in phytoagar during cultivation and to provide a basis for further sample treatment with the aim of Se isotope determinations. Because of a relative recovery of 100% (Figure 25, Table 6) and the improbable sorption reported by Davies et al. (2010) no significant isotope effects were expected. Residual TOC was with 112.4 (±51.5) mg L^{-1} (n=3) (6 % of initial C_{org}) significant, but samples derived from inorganic Se species supplementation did not include Se_{org} as it applies for plants. For SeMet supplied cultivation batches, further organic destruction might be necessary to gain highly precise isotope values (raw data available in Appendix IV, Table IV-5).

Table 6: Se recoveries dependent from Se concentration and species added using digestion after Kopp (1999)* and vacuum filtration** (raw data available in Appendix IV, Table IV-3)

Se species	**selenate**			**selenite**			**SeMet**			
Se concentration [µg L^{-1}]	**100**	**500**	**1000**	**100**	**500**	**1000**	**100**	**500**	**1000**	**average**
Se recovery [%]*	77.5 ±9.2	84.8 ±7.6	81.2 ±4.0	58.3 ±1.5	76.0 ±3.0	72.0 ±4.1	76.5 ±0.6	70.3 ±4.5	71.8 ±4.9	74.3 ±5.5
n*	2	2	2	2	2	2	2	2	2	18
Se recovery [%]**	103 ±0.6	105 ±0.6	108 ±0.2	103 ±0.5	109 ±0.3	103 ±0.9	101 ±2.5	102 ±3.6	97.7 ±1.9	103 ±2.6
n**	3	3	3	3	3	3	5	5	5	33

The data presented shows that vacuum filtration of phytoagar is a very successful method for the determination of Se concentrations and species, because it provides full recovery, only small dependency on source species as well as species conservation. It is more reliable than digestion after Kopp (1999) that is characterized by high Se losses and bad reproducibility.

Additionally, vacuum filtration has a huge potential for the measurements of other trace elements or molecules in phytoagar as long as they do not tend to interact with its molecular grid. The applicability of purification methods on phytoagar extracts (chapter 4.4) will eventually show if vacuum filtration can be suitable for the treatment of samples with regard towards Se isotope determinations according to the analytical method applied (chapter 4.5).

4.3 Plant treatment

Plant samples are challenging concerning destruction and homogenization due to their firm cell structure. The tissue mainly consists of cellulose, lignin, pectin, hemicellulose and water soluble components with varying fractions depending on the plant part (root, leave, stem). The composition of those compounds defines the degree of polymerization and therefore determines the stability of the tissue (Franck, 2005). Furthermore the ratio of organically bound Se significantly differs depending on the plant part. To gain reliable, valid results, organic compounds must be fully and equally destroyed within the entire plant. Finding an adequate method for minimization and homogenization is therefore a mandatory prerequisite for further treatment. Reduction of material losses during this procedure will be another requirement as the *Minimum Parameter* setups (chapter 5.1) provide low absolute sample amounts (Appendix IV, Table IV-14). Building on this, a digestion method must be implemented that ensures full recovery and full organic destruction, because valid preparation and analytics demand Se that is totally available as an inorganic compound (chapters 3.2 and 4.4). Additionally, organic fractions in general might disturb those processes because of their selective interaction with Se and other sample compounds (Wasilewska et al., 2002; Zsolnay, 2003).
After harvesting the cultivated rice plants (chapter 5.1), the fresh plant material was washed with H_2O (millipore) and transferred to a 2 mL Eppendorf cup containing a 5 mm diameter stainless steel bead. The cups were frozen with liquid nitrogen and directly transferred into the Tissue Lyzer, an

electric mill designed for small amounts of plant tissue (Qiagen, Venlo, Netherlands). After milling at 30 Hz for 60-120 s – until the tissue was powdered – the powder was lyophilized (Alpha 1-4 Freeze Dryer, Martin Christ, Osterode, Germany) for 24 h in the open Eppendorf cups and mortared to a homogeneous powder afterwards.

Bell et al. (1992) firstly described a widely applied, microwave based plant digestion method (Table 7). It was proved to be insufficient for Se isotope analytics due to high organic residuals and high contamination risks. Therefore a digestion method designed for Se determinations in organic samples was implemented, which was based on a closed microwave system with two separated reaction chambers, the microwave beaker and a quartz vessel included therein (START 1500, MLS, Leutkirch, Germany) (Figure 26). The sample with 2 mL concentrated HNO_3 was located inside the quartz vessel that had a loose lid on top, whereas 7 mL of diluted (8.6 %) H_2O_2 was poured into the beaker (Figure 26 (a)). It was closed, placed into a high pressure container, then heated up in the microwave slowly to 240°C and held for 20 min (Table 7).

Table 7: Technical parameters applied in tests of plant digestion methods according to Bell et al. (1992) and Kopp (1999) using the same sample amount of 0.1 g plant tissue

Step	**Bell et al. (1992)**	**Kopp (1999)**
Sample container	PFA	quartz
Acid addition	3 mL 65 % HNO_3 + 2 mL 15 % H_2O_2 inside PFA beaker	2 mL 65 % HNO_3 inside + 2 mL 8.6 % H_2O_2 outside quartz vessel
Heating step 1	3 min at 75°C (600 W)	1.5 min at 65°C (500W)
Heating step 2	8 min at 130°C (700 W)	4.5 min at 130°C (500W)
Heating step 3	10 min at 210°C (1000W)	3.5 min at 210°C (1000W)
Heating step 4	12 min at 220°C (1000W)	3.5 min at 240°C (1000W)
Heating step 5	n/a	20 min at 240°C (1000W)
Ventilation	45 min	30 min
Cooling down	minimum 12 hours	minimum 12 hours

The same procedure was used to clean the equipment between two digestion batches. During digestion the plant tissue was oxidized by HNO_3. CO_2 and NO were released to the gas phase within the quartz vessel. High pressure enabled CO_2 and NO to leave the vessel by lifting the lid, which reduced the CO_2 partial pressure inside and therefore increased mineralization rates. NO was kept by the surrounding H_2O_2 in the PFA beaker and oxidized to HNO_3 (Dr. Gernot Kopp, personal comment). This transformation of NO from gaseous to aqueous phase lowers the partial pressure in the PFA beaker and enables the use of higher temperatures without busting the technical pressure limits. Whereas CO_2 and NO leave the quartz vessel at a critical pressure point and therefore reduce the pressure inside, the heavier volatile Se compounds remain in the quartz vessel and dissolve into the digest again at cooling (Figure 26 (b)). Further advantages of this method compared to Bell et al. (1992) are the lower blank by using quartz instead of porous PFA in sample contact as well as the lower amount of HNO_3 reducing blanks and difficulties at further sample preparation steps.

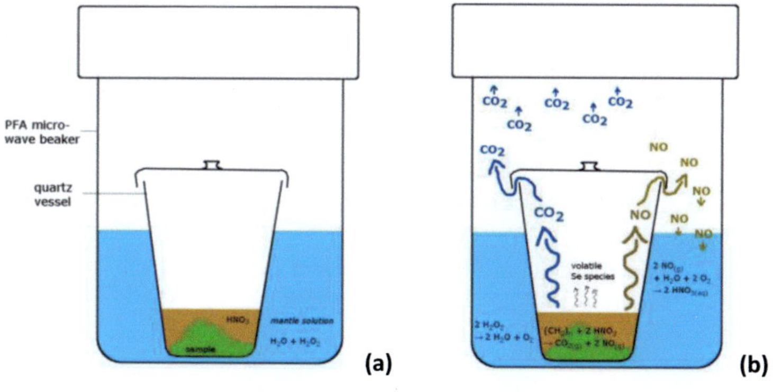

Figure 26: Implemented digestion method after Kopp (1999) (a) preparation of digestion beakers. (b) Transport and transformation processes during digestion.

The theoretical concept of this digestion procedure promises higher mineralization rates, higher Se recoveries and lower blanks than comparable methods, but it was not systematically monitored on these factors yet (Kopp, 1999; Dr. Gernot Kopp, personal comment). In order to test reliability and validity, digestions with certified reference material (NISTSRM1567a) were performed. This reference material had a wheat flour matrix, which is closest possible to rice plant tissue used within this study. According to Tsai and Jiang (2011) the organic Se fraction of NISTSRM1567a was very high (0.9 ppm, 70-100 %), which makes it highly suitable for digestion efficiency tests. For each sample 0.1 g of NISTSRM1567a was weighted and digested according to Bell et al. (1992) and Kopp (1999) to evaluate potential differences. Se concentrations were determined in the solid plant tissue using energy dispersive x-ray spectrometry (EDX) and in the digest using ICP-MS (chapter 3.1.1). Those concentrations were then compared. The initial organic carbon content C_{org} in the solid plant tissue was measured as total carbon (C) (CSA 5003, Leybold-Heraeus) assuming that the entire C in the plant tissue was organically bound. The residual C_{org} in the digest was measured as TOC (chapter 3.1.2). Se recoveries, C_{org} residuals and mineralization rates were calculated according to Equations (78) to (80) from the absolute amounts of Se [µg] (a(Se)) and organic C [mg] (a(C_{org})).

$$Se\ recovery\ [\%] = \frac{a(Se)_{digest}}{a(Se)_{initial}} * 100 \tag{78}$$

$$C_{org}\ residuals\ [\%] = \left(\frac{a(C_{org})_{digest}}{a(C_{org})_{initial}}\right) * 100 \tag{79}$$

$$mineralization\ rate\ [\%] = \left(1 - \frac{a(C_{org})_{digest}}{a(C_{org})_{initial}}\right) * 100 \tag{80}$$

These results were compared to systematic digestion tests on organic destruction performed by Wasilewska et al. (2002). Afterwards cultivated and minimized plants were digested according to Bell et al. (1992) and Kopp (1999) and the same parameters were measured to test the transferability to the target samples. The influence of the minimization procedure described above on the digestion efficiency was quantified as well.

Due to the wide error range of the certified value of NISTSRM 1567a, the used batch was again analyzed on Se and measured as 1.25 (±0.15) ppm (n=6) with EDX. The C content was 42.76 (±0.69) % (n=6), which was assumed to be completely organically bound (raw data available in Appendix IV, Tables IV-4 and 5).

Using the digestion method by Bell et al. (1992), the average Se yield for NISTSRM1567a was 0.94 (±0.06) ppm (n=31), which is on the lower boundary of the certified range. Relatively high Se residuals in the microwave beakers after digestion were expected. That is why they were determined by cleaning the beakers with boiling 7.5M HNO_3 for 1 h, rinsing with H_2O, boiling again with 7.5M HNO_3 for 1 h and measuring the Se concentration in the second solution. Unreproducible Se residuals of up to 1.31 µg Se (131 µg L^{-1}) were detected. The residual TOC concentrations were 334.73 (±37.33) mg L^{-1} (n=9), which equals 7.83 % of the initial carbon content. As organically bound Se cannot be considered for further sample treatment and correction mechanisms (chapters 3.5 and 4.3), 8 % of the entire Se would be neglected within isotope composition measurements, if plant samples were digested like this. A spot test on Se species composition according to Bird et al. (1997) (chapter 3.1.4) even detected an organic Se fraction of 15.7 % being twice as high (Appendix IV, Tables IV-5 and 12). The isotope composition highly depends on the oxidation state and the molecular form (Johnson, 2004), which is why significant isotopic bias could result from such a high organic Se fraction. Additionally, analytic disturbances by organic molecules are likely (Wasilewska et al., 2002; Zsolnay, 2003). Digestion after Kopp (1999) resulted in Se concentrations around 1.01 (±0.08) ppm (n=9), which is in the centre of the certified Se range. All Se blanks measured were below 1 µg L^{-1}, probably because the quartz vessels were easier to clean and the digest did not get in contact with the potentially Se contaminated microwave beaker. TOC residuals were 30.51 (±18.91) mg L^{-1} (n=9) corresponding to 0.69 % of the initial C_{org}, which was quantified to 43.75 (±0.12) % (n=6). The variability was relatively high, but there was no sample with TOC residuals of more than 1.5 %. Wasilewska et al. (2002), who tested four digestion procedures for organic material, one high pressure asher and three microwave based ones, characterized a TOC residual of <2 % in digests as fully mineralized. This was only reached for one of the methods checked at the same temperature of 240°C. Residual TOC was tested on the target plant samples with and without performing the minimization process described in chapter 4.3, using the digestion method after Kopp (1999). Without complete minimization a residual TOC of 178.12 (±49.01) mg L^{-1} (n=9) (4.16 %) was measured. Results with minimization were 57.01 (±28.12) mg L^{-1} (n=9) (1.29 %). This significant difference proves that the minimization process is necessary and reduces the potential error by further 68 % (raw data available in Appendix IV, Tables IV-4 and 5). The presented data shows that using digestion after Kopp (1999) reduces the potential error by 91 % related to Bell et al. (1992) and only when regarding residual TOC. Higher Se recoveries and lower Se blanks are further advantages of Kopp (1999). Therefore, this digestion procedure combined with thorough plant tissue minimization is an effective method for the reduction of organic residuals and the minimization of Se losses, making it suitable as sample treatment procedure for Se isotope determinations in plants.

4.4 Purification

As described in chapter 3.2.2 there is a variety of disturbing sample components, especially isobaric interferences on Se masses. Therefore the target samples must be purified prior to Se isotope analytics. Three methods are predominantly described in literature and applied mainly for geological samples, but no systematic tests on their purification efficiency and limitations were published. To provide a general aid finding the adequate purification method depending on the sample matrix and to choose the suitable one for plant and phytoagar, those three methods were carried out and monitored according to their process steps. High concentrated multi-element standard solutions as well as plant and phytoagar samples were used. Se recovery, residual elements and residual organic compounds (as TOC) were determined as quality indicators.

4.4.1 Methodical setup

The three methods were firstly described by Ellis et al. (2003) (modified) (referred to as method (A)), Elwaer and Hintelmann (2008c) (modified) (referred to as method (B)) and Clark and Johnson (2008) (referred to as method (C)). Methods (A) and (B) are based on selective Se retention in flow-through columns (5 mL Minicolumns, Spectrum Labs) packed with different materials: (A) commercial anion exchange resin (AG1-X8, 100-200 µm dry mesh size, BioRad) and (B) in-house produced thiol activated cellulose powder (TCP). In method (C) gaseous Se hydrides (H_2Se) are generated from the liquid sample and the gas phase is separately trapped.

Method (A) – anion exchange

The anion exchange resin AG1-X8 – referred to as method (A) – consists of a styrene divinylbenzene copolymer lattice with quaternary ammonium functional groups (fixed phase), which is positively charged, saturated with chloride counter anions (mobile phase) (BioRad, 2011) (Figure 27).

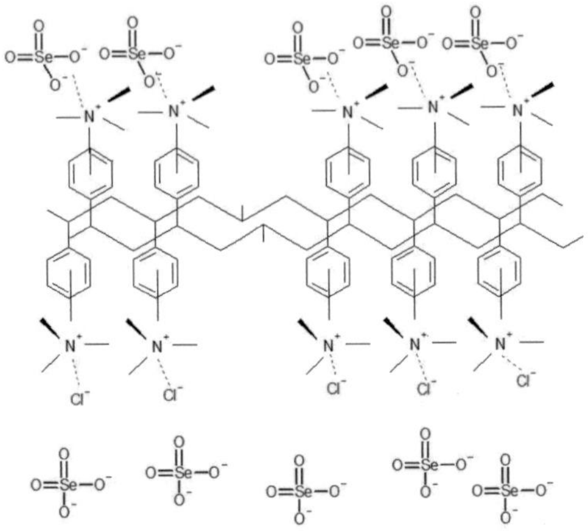

Figure 27: Molecular structure and retention/desorption mechanism in purification method (A) – selective outerspheric sorption of selenate (SeO_4^{2-}) ions and subsequent desorption by more competitive chloride (Cl^-) ions.

The use of AG1-X8 is applied among others for sample preparation of Fe, Cr and Se isotope determinations (Schoenberg and von Blanckenburg, 2005; Zink et al., 2010; Ellis et al., 2003; Clark and Johnson, 2008; Schilling et al., 2011a+b) as well as for the separation of Cr, S, As and Se species in aqueous solutions (Zink et al., 2010; Druschel et al., 2003; Kim, 2001; Pohl and Prusisz, 2004; Ellis et al., 2003). However, none of the studies systematically tested it for purification of Se samples. Chloride anions (Cl^-) have a particular affinity to the resin's surface at low pH value, whereas at neutral pH, Se oxyanions sorb to the positively charged resin's surface via hydrogen bridges (BioRad, 2011). Thereby the affinity of selenate is higher than that of selenite, because selenate is totally available as double charged SeO_4^{2-} at neutral pH, whereas selenite is up to 90 % available as single charged $HSeO_3^-$ (Figure 28 (a)). Se can easily be desorbed by adding HCl providing an acid environment and Cl^- for exchange with SeO_4^{2-} at the sorption spaces. Figure 27 illustrates the resin's organic structure as well as the sorption and remobilization mechanisms.

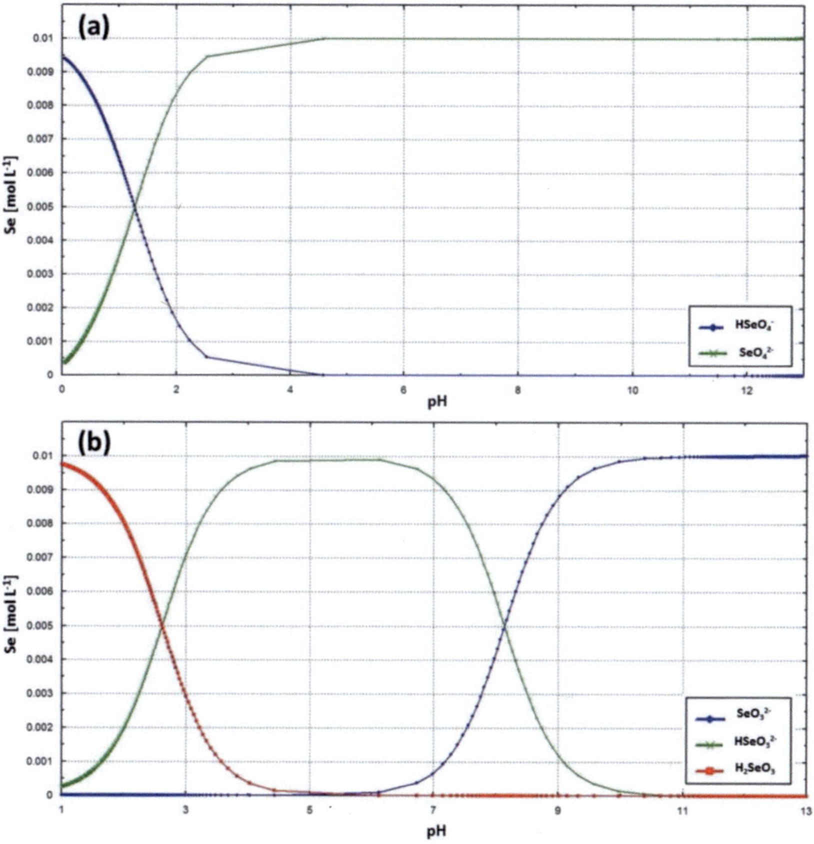

Figure 28: Abundances of selenate (a) and selenite (b) species dependent on the pH value (calculated with PhreeqC, wateq4f data base for a 0.01M matrix free Na_2SeO_4 (a) respectively Na_2SeO_3 (b) solution).

About 50 g of the resin powder was transferred to a PFA bottle and washed successively with 100 mL methanol, 100 mL 1M NaOH and 100 mL 1M HCl (Dr. Kathrin Schilling, personal comment). In analogy to Ellis et al. (2003), 1.2 mL of resin suspension were filled into a 5 mL minicolumn (SpectrumLabs), activated by passing 10 mL 6M HCl and neutralized by passing

H_2O (millipore). The pH was monitored with indicator strips (VWR). The sample was brought into a neutral H_2O matrix. 100 µL of a 0.25 mM $K_2S_2O_8$ solution were added to the sample and heated up for 60 min at 120°C to completely oxidize Se to SeO_4^{2-}. After cooling down, it was added to the column. Subsequently, 10 mL 0.1M HCl were added followed by H_2O (millipore) to remove retained matrix residuals (wash step). 5 mL 6M HCl were added to remobilize selenate via exchange with Cl^- (Se extract). As this procedure yielded improvable results in the first place, several modifications were separately tested to improve the success of the method:

(a) after pouring the resin it was compressed with a stirring rod to provide continuous flow rates and more homogeneous chemical conditions in all columns
(b) the sample was taken up in 0.1M HCl to create slightly reducing and acidic conditions in order to keep competing anions in solution and therefore avoid the blocking of sorption spaces; furthermore the wash step was skipped to avoid the removal of Se prior to extraction
(c) within the wash step, 0.1M HCl was replaced with H_2O to avoid the removal of Se prior to extraction

The original method by Ellis et al. (2003) and three variations were performed with plant digests in small test series at four columns each, Variation I by only applying modification (a), Variation II by only applying modification (b) and Variation III by applying modifications (a) and (c) combined. After testing and evaluating those variations on Se recovery and residual matrix elements (chapters 4.3.3.2 and 4.3.3.3), Variation III was implemented as the most successful one and all experiments presented were carried out according to it, as described here.
The 1.2 mL of the resin was compressed with a stirring rod after pouring into the column. Sample preparation was done in analogy to Ellis et al. (2003) as described above. By adding the sample to the column selenate was retained whereas most matrix elements passed through. The eluate derived from

sample addition was kept for Se and matrix concentration analysis (*sample eluate*). 20 mL H_2O were added to wash out the matrix elements that were retained within the column. The eluate was kept for concentration analysis as well (*wash eluate*). In a next step, 5 mL 6M HCl in five steps of 1 mL were added to desorb selenate via exchange with Cl^-, while selenate was remobilized and collected separately. This eluate was kept for concentration and, for selected samples, for isotope analysis (*Se extract*). The aliquot provided for Se isotope analytics was evaporated to dryness at 70°C and diluted to 2M HCl. Figure 29 illustrates the four steps of this procedure. The flow rates depended on the individual permeability of the resin and were on average 0.52 (±0.08) mL min^{-1} (n=10). No obvious dependence of Se yield and matrix residuals on flow rates was detected. According to BioRad (2011) the capacity is 1.44 meq per column (1.2 mL resin), which equals 56.8 mg Se, available as SeO_4^{2-}. A detailed instruction on purification according to method (A) is given in Appendix III.

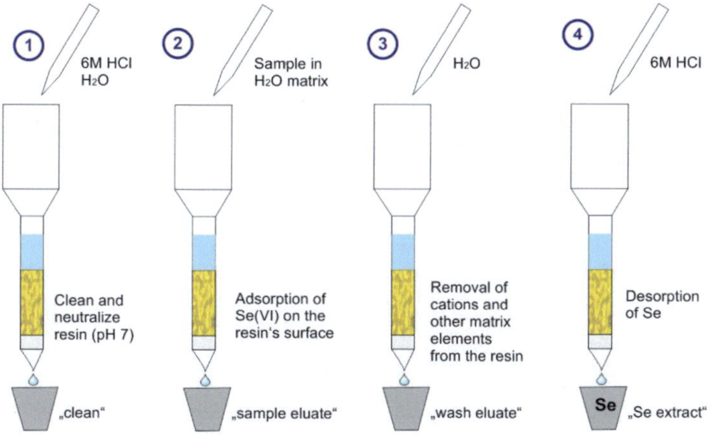

Figure 29: The four steps of purification according to method (A) – optimized from Ellis et al. (2003): 1) Activation of packing material in the column, 2) retention of Se in the packing material, 3) removal of matrix elements and 4) extraction of Se from the packing material.

Method (B) – thiol retention

Method (B) was based on selective Se retention in thiol activated cellulose powder (TCP) and set up by Elwaer and Hintelmann (2008c) for Se isotope determinations. It makes use of the exceptionally high affinity of Se to thiol groups (-SH) to form covalent bonds, which are broken by HNO_3 afterwards. TCP was produced in the laboratory by controlled and catalyzed reactions of cellulose powder with thioglycolic acid (Figure 30).

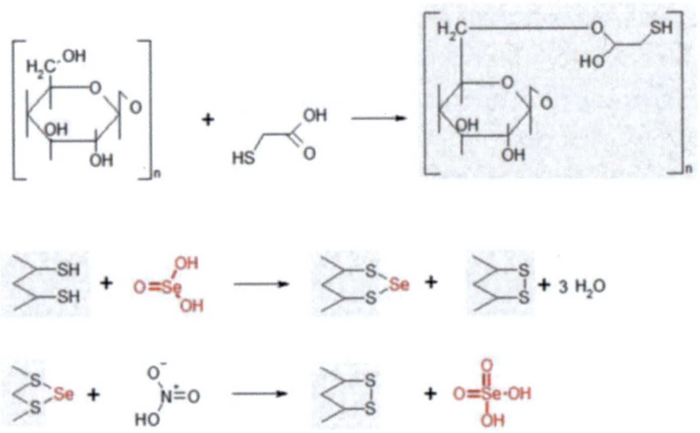

Figure 30: Purification mechanisms in method (B): Activation of cellulose powder with thioglycolic acid (upper), binding of selenite to thiol groups as innerspheric complex (central), extraction of Se from TCP with HNO_3 (lower) (modified from Elwaer and Hintelmann, 2008c).

This method was slightly modified with regard to sample matrices, analytic Se signal strength, qualitative reliability in TCP production and technical effort by the following measures:

(a) The temperature at TCP production was reduced from 60°C to 55°C. It turned out empirically that quality and drying ability of the TCP increased with the lower temperature.

(b) The reduction time of the samples before addition to the columns was increased to ensure full reduction in the presence of high matrix element concentrations and residual HNO_3 from digestion. Thereby, the temperature was declined to avoid volatile Se losses.
(c) The flow rate was gravity driven instead of being operated by a vacuum system. It slowed down the speed by factor 4 which provided a longer reaction time of the sample with the TCP along with a lower technical effort and decreased risk of contamination.
(d) The amount of concentrated HNO_3 in the extraction step was increased by factor 10 to 1 mL (total) in order to raise the Se recovery. The Se signal suppression by HNO_3 reported by Elwaer and Hintelmann (2008c) was experienced too, but avoided by evaporating the samples at 70°C and taking them up in the matrix needed for analytics.

Those modifications were included into the implemented method based on Elwaer and Hintelmann (2008c) and used for all experiments as described here.

For preparation of TCP, 5 g cellulose powder (SigmaAldrich, SigmaCell Type 20) were weighted into a 250 mL PFA bottle and 30 mL conc. thioglycolic acid (98 %) (AppliChem), 15 mL acetic anhydride (98 %), 10 mL acetic acid (96 %) and 0.5 mL sulfuric acid (98 %) (Merck) were added. The mixture was shaken for 30 min, heated up in a water bath (E30U, Dinkelberg) for 24 h at 55°C, shaken again for 30 min and heated up again for 24 h at 55°C. Then it was washed with H_2O, filtered and dried at room temperature. The dry material was mortared to a fine homogeneous powder.

The evaporated sample was diluted to 1.7 mL 6M HCl and fully reduced to selenite in a closed beaker by heating up on a hotplate for 90 min at 80°C. The sample was cooled down and diluted to 10 mL 1M HCl. Under acidic conditions selenite is totally available as H_2SeO_3 (Figure 28 (b)) and therefore affine to thiol binding (Figure 30) (Elwaer and Hintelmann, 2008c). 0.1 g TCP was filled into each column. The TCP was cleaned and conditioned by passing 2*2 mL H_2O, 2 mL 6M HCl and 2 mL 1M HCl. Afterwards, the sample was added to the column, inducing selenite to form covalent bonds to the reactive thiol (-SH) groups on the TCP surface (Figure 30). The eluate was kept for concentration analysis (*sample eluate*). 2 mL 6M HCl, 2 mL H_2O and 2 mL 1M HCl were added to the column to wash out the thiol affine matrix elements by forming chloride complexes. This eluate was kept for concentration analysis as well (*wash eluate*). After the column was completely emptied, the Se containing TCP was transferred to a 50 mL centrifuge tube (VWR) and 500 µL concentrated HNO_3 and 500 µL H_2O were added. The tube was closed and heated up in a boiling water bath (100°C) for 20 min. After cooling down, 3 mL H_2O were added; the tube was shaken and centrifuged for 10 min at 4000 RPM (Rotofix 32A, Hettich). The supernatant was kept in a separated PFA beaker, the process was repeated with the residual powder and both supernatants were combined. This sample was analyzed on its Se concentration and, for selected samples, on its Se isotope composition (*Se extract*). The aliquot for isotope determinations was evaporated at 70°C, repeatedly oxidized with 1 mL mixture of concentrated H_2O_2 and 0.5M HNO_3 (1:10) until the TCP was completely dissolved (Zhu et al., 2008b), evaporated to near dryness afterwards and diluted to 2M HCl. Figure 31 illustrates the purification steps within method (B). The flow rates were on average 0.18 (±0.01) mL min^{-1} (n=10). No influence of the individual flow rate on the results was detected. According to Elwaer and Hintelmann (2008c) the capacity was 25 µg Se per 0.1 g TCP. A detailed instruction on purification according to method (B) is given in Appendix III.

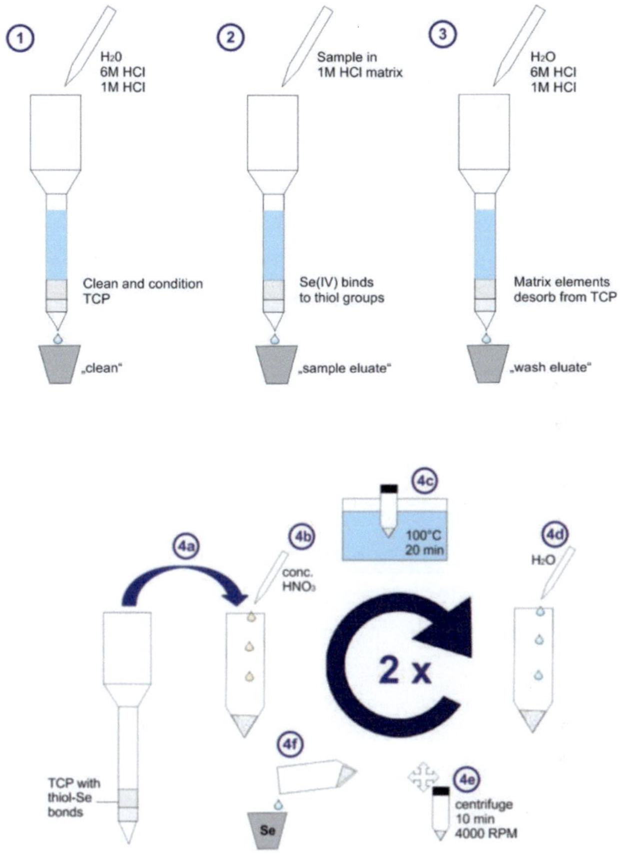

Figure 31: The four steps of purification according to method (B) – optimized from Elwaer and Hintelmann (2008c): 1) Activation of packing material in the column, 2) retention of Se in the packing material, 3) removal of matrix elements and 4) extraction of Se from the packing material by a) transferring to centrifuge tube, b) addition of HNO_3, c) boiling in water bath, d) addition of H_2O, e) centrifugation, f) removal of supernatant (afterwards repetition of b) to f)).

Method (C) – hydride separation

Method (C) makes use of the hydride generating property of Se. Like few other elements (e.g. As, Ge, Sb, Te) Se forms volatile H_2Se molecules in its lowest oxidation state Se(-II) if H and reducing agents are sufficiently present. This effect is widely applied in analytical chemistry (Campbell, 1992) and was, with some modifications, performed in analogy to on-line HG as sample introduction for Se isotope analytics (chapter 3.2.6). By adding strong reducing agents and H ($NaBH_4$) to the Se containing sample, Se transfers into the gas phase, whereas non-hydride generating elements remain in the liquid phase. A gas-liquid separator enables a discharge of the gas phase, which is then introduced into a strongly oxidizing alkaline trapping solution to transform Se into soluble aqueous anions again. Heating up the trapping solution leads to a full transformation to Se(VI) (Figure 32). A subsequent anion exchange according to method (A) (chapter 4.4.1) separates Se from other by-trapped hydride generating elements and removes the alkaline matrix (Clark and Johnson, 2008).

Method (C) was applied in analogy to Clark and Johnson (2008). Before HG, HNO_3 digests, phytoagar filtrates or H_2O matrices were diluted to 4M HCl. Samples containing hydrofluoric acid (HF) residuals should be evaporated in advance, because unlike HNO_3 HF might inhibit HG (Welz and Melcher, 1981; Welz, 1983). The samples were reduced by heating up at 80°C for 90 min to convert all Se into Se(IV) as only Se(IV) leads to HG. After cooling down they were diluted to 2M HCl by adding H_2O in the required amount. 2M HCl keeps Se(IV) stable and is a suitable reagent for HG (Clark and Johnson, 2008) (chapter 3.2.6). All samples were taken up by the hydride system using a peristaltic pump (80 RPM) with an uptake rate of 4.1 mL min^{-1}. The $NaBH_4$ solution, which consists of 2.4 g L^{-1} $NaBH_4$ granulates (Merck) and 4 g L^{-1} NaOH (Merck), was continuously supplied with an uptake rate of 2.8 mL min^{-1}. Both solutions were mixed in a tubing loop, which gave sufficient time for the reduction, HG and phase separation (chapter 3.2.6, Equation (11)). Both gaseous and liquid phase were introduced into a gas-liquid separator (FIAS 400, Perkin Elmer) in which Ar gas was added in order to push the gaseous H_2Se upwards and transport it into the trapping solution (Figure 32). The liquid sample matrix was discharged from the gas-liquid separator by pumping with 5.6 mL min^{-1}. After taking up the sample solution, the probe was placed into 2M HCl, which was taken up for 60 sec (4.1 mL) in order to recover all Se retained within the hydride system. After taking aliquots for concentration analyses, the Se containing traps were heated up at 80°C for 60 min to convert all Se into Se(VI) and afterwards cooled down and purified in analogy to method (A) procedure (chapter 4.4.1, Figure 29). *Sample, wash* and *Se extract eluates* deriving from (A) application were kept for analysis to trace matrix removal. *Se extract eluates* of selected samples were used for Se isotope analytics. A detailed instruction on purification according to method (C) is provided in Appendix III.

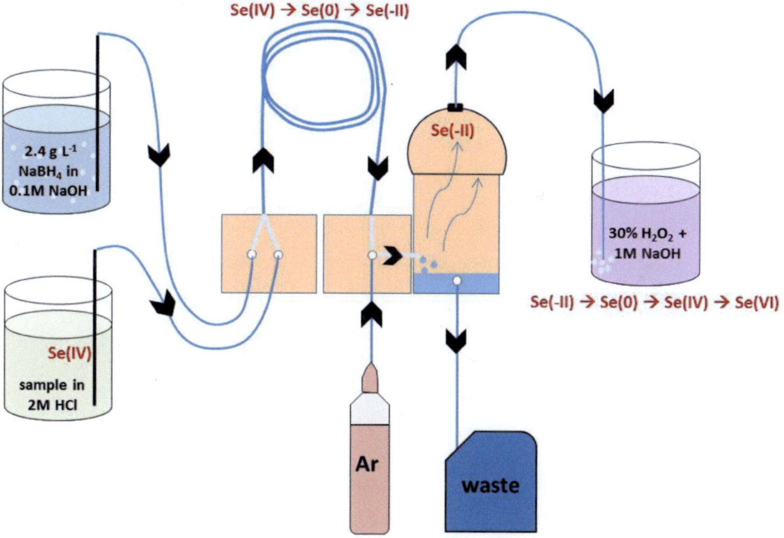

Figure 32: Scheme of purification method (C) - HG and trapping for the separation of Se from the sample matrix.

4.4.2 Test matrices

Various sample types and matrices were tested using purification methods (A)-(C). Punjab plants were digested after Bell et al. (1992), whereas cultivated plants and phytoagar were prepared as described in chapters 4.1 and 4.2. SGR-1 reference (USGS) was organic and Se rich shale (25 % TOC, 3.5 ppm Se (USGS)), which was tested on Se isotope composition by several studies and, among others, used for method validation in this study (chapter 4.5.3). It was digested with a HF-HNO_3-$HClO_4$ approach modified from Layton-Matthews et al. (2013), Pogge von Strandmann et al. (2014) and Zhu et al. (2014). 200 mg of SGR-1 were weighted in a PFA beaker and 3 mL conc. HNO_3 was added. The closed beaker was placed on a heating plate and heated up at 80°C for 16 h to destroy organic compounds. After cooling down, 3 mL conc. HF and 0.5 mL conc. $HClO_4$ were added and the beaker was heated up at 80°C for another 76 h. The digest was evaporated to ~100 µL at 70°C, 3 mL concentrated HNO_3, 3 mL concentrated HF and 0.5 mL concentrated $HClO_4$ were added and evaporated to ~100 µL again. This step was repeated three times. Afterwards precipitated fluoride compounds were solved again by adding 3 mL conc. HNO_3. The sample was evaporated to near dryness at 70°C. Table 8 lists the sample matrices, their characteristics and the availability of concentration measurements (Na, Mg, Al, (P), Ca, Cr, Fe, Co, Ni, Cu, Zn, Ge, As, Se) in aliquots taken during purification procedure. The bold samples were additionally measured on Se isotope composition (chapter 4.5).

Table 8: Sample types measured in purification steps within methods (A), (B) and (C) (* multielement = Ag, Al, B, Ba, Bi, Ca, Cd, Co, Cr, Cu, Fe, Ga, In, K, Li, Mg, Mn, Na, Ni, Pb, Sr, Tl and Zn; ** Na (3.57 (±3.14) µg), Mg (9.71 (±4.24) µg), Al (4.74 (±5.70) µg), P (9.13 (±3.29) µg), Ca (27.3 (±16.4) µg), Fe (3.06 (±3.56) µg), Cr (20.4 (±15.9) ng), Co (1.48 (±1.59) ng), Ni (13.4 (±10.3) ng), Cu (28.5 (±16.0) ng), Zn (109 (±67.4) ng), Ge (0.72 (±0.62) ng), As (2.03 (±2.30) ng) and Se 0.81 (±0.03) (individual concentrations of the initial plant tissue from Punjab Plants are provided in Appendix IV – Table IV-6))

Sample type	**Origin**	**Hydride trapping**	**Sample eluate**	**Wash eluate**	**Extract eluate**
Pure Se	Roth AAS Se standard	(C)	-	-	(A), (B), (C)
Multielement ICP standard	100 µg multielement* + 10 µg As, Ge and Se (Roth, Alfa Aeser)	(C)	(A), (B), (C)	(A), (B), (C)	(A), (B), (C)
Punjab Plants	Se rich wheat crops and Brassica**, digest aliquots (provided by E. Eiche)	-	(A), (B), (C)	(A), (B), (C)	(A), (B), (C)
Extracted phytoagar	containing long-chain hydrocarbons, but poor in matrix elements	(C)	(C)	(C)	**(A), (B), (C)**
Cultivated Plants	16 days old rice plants in *Minimum Parameter* approach, poor in matrix elements	(C)	(C)	(C)	**(A), (B), (C)**
Shale digest	SGR1 reference material (Green River Shale) (USGS)	(C)	(C)	(C)	**(C)**

Pure Se was chosen to examine the principal functionality and the extent of potential matrix influence on the Se recovery. With the ICP multi-element standard the matrix removal pathways as well as the influence of main elements and critical metals were intended to trace using exceptionally high concentrations. Punjab plants were used in order to give an impression in how far the methods were applicable to plant tissue that was characterized by residual organic molecules and naturally ubiquitary anions such as phosphate, sulphate and nitrate. Phytoagar and cultivated plants derived from the cultivation experiments were taken to test and monitor the efficiency for target samples within this study. SGR-1 reference was included to have an additional environmental sample material in order to improve information on universality of the methods. Furthermore this standard was measured on Se isotope composition by other research groups and was therefore used within the validation process (chapter 4.5.3).

4.4.3 Data processing

All element concentrations were corrected on process blanks, which were included in any purification batch together with maximum nine samples. Residual matrix element concentrations are given without blank correction to assess the suitability of method and laboratory conditions for the acceptable blank ranges for Se isotope analytics. Se recovery as well as removal pathways were determined using a mass balance model approach. For that purpose all element concentrations (c) given in [µg L^{-1}] were calculated to absolute amounts [µg] (a) regarding sample volumes (V) [L] according to Equation (81).

$$a = c * V \tag{81}$$

Se recovery [%] was calculated using the Se amount added ($a(Se)_{added}$), which equals the amount in the initial sample, and the Se amount in the Se extract phase ($a(Se)_{extract}$) according to Equation (82).

$$Se\ recovery = \frac{a(Se)_{extract}}{a(Se)_{added}} * 100 \tag{82}$$

In analogy to the Se recovery, the fraction removed from the column (removal [%]) was calculated using the element amounts in the *sample eluate* (a_{sample}), *wash eluate* (a_{wash}) and *Se extract* ($a_{extract}$) phase and the element amounts added (in the initial sample) (a_{added}) according to Equations (83)-(85).

$$Removal\ by\ sample\ addition\ [\%] = \frac{a_{sample}}{a_{added}} * 100 \quad (83)$$

$$Removal\ by\ wash\ [\%] = \frac{a_{wash}}{a_{added}} * 100 \quad (84)$$

$$Removal\ by\ Se\ extraction\ [\%] = \frac{a_{extract}}{a_{added}} * 100 \quad (85)$$

The deficit [%] that describes the amount retained in the column after extraction or lost during the process was calculated as a remainder using sample, wash and extract phase values compared to the amount added (Equation (86)).

$$deficit\ [\%] = \frac{(a_{sample}+a_{wash}+a_{extract})}{a_{added}} \quad (86)$$

4.4.4 Removal pathways of critical elements

The results of the ICP multi-element standard samples were used to examine removal pathways of individual matrix elements. They contained main and trace elements in equally high amounts and therefore show interference potentials of single elements and their interactions within the system.

Method (A) – anion exchange

Figure 33 shows the removal pathways of the elements using the ICP multi-element standard and method (A).

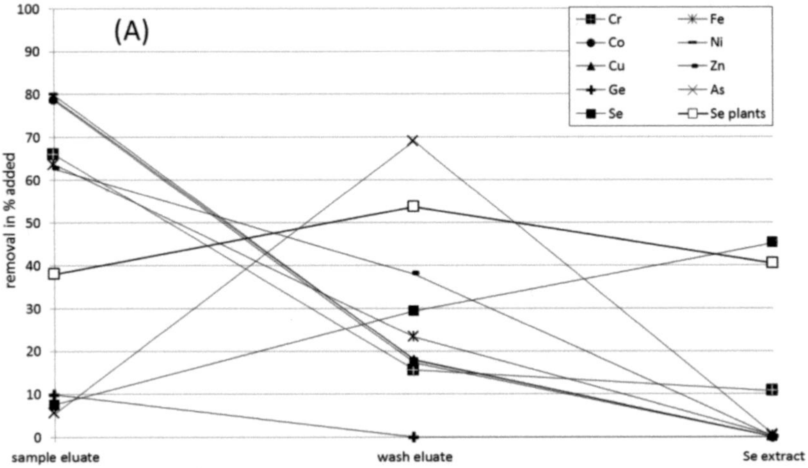

Figure 33: Matrix element and Se ratios (related to initial Se in sample) determined in each step eluate of method (A) derived from the ICP multi-element standard and for Se only for plant digests (raw data available in Appendix IV, Table IV-6).

The major load of 63-86 % of all elements determined – except As, Ge and Se – was removed in the *sample eluate* step indicating that it stayed in solution and did not significantly interact with the resin. The residual fraction was washed out in the second step only using H_2O to rinse the transport channels, which indicates that this fraction was not bound to the column, at most slightly adsorbed or retained in dead end pores or slower transport channels. There was no or negligible low matrix element fraction in the *Se extract* solution except for the one of Cr (10 %). It was assumed that the similar geochemical properties of chromate and selenate – with higher affinity of Se to the resin - cause the relatively high residuals. The material is therefore suitable and applied for Cr separation as well (Zink et al., 2010), if

Se concentrations are low. Arsenic was found by 5 % in the sample eluate indicating that the most oxidized As species, arsenate ($H_2AsO_4^-$ and $HAsO_4^{2-}$ at neutral pH), adsorbed to the resin due to its double negative charge analogous to selenate. Therefore AG1-X8 is suitable for arsenate applications as well and applied by several studies (e.g. Kim, 2001; Pohl and Prusicz, 2004). However, As was removed by almost 70 % (7 µg) in the wash step indicating that there was another removal mechanism than anion exchange. A high amount of Fe (23 %, 23 µg) was found in the wash eluate as well, which in oxidizing environment tend to form ferric arsenate complexes (Gao et al., 2013). Due to their mainly positive charge they do not adsorb to the resin, but are longer retained in the column because of larger molecule size by precipitation as nanoparticles. The high deficits of Fe (13 %, 13 µg) and As (15 %, 1.5 µg) indicate longer retention within the packing material. Ge was removed by 10 % in the *sample eluate*, but no more in the wash and extraction phase, which indicates that the major fraction was still retained in the column. Pokrovsky et al. (2006) reported the extensive co-precipitation of Ge with iron oxy(hydr)oxides formed during Fe(II) oxidation or by Fe(III) hydrolysis in neutral solutions. This led to the formation of high Ge incorporations into solid Fe phases. Method (A) purification includes similar conditions, which is why huge amounts of Ge might be retained as precipitates in dead end pores or tiny inactive flow channels within the column. Method (A) might cause problems if samples with high Cr, As and potentially Ge concentrations are used, as all of them are analytically critical elements (chapter 3.2.2). Plants usually contain all three elements only at trace level if they do not derive from contaminated sites.

Se tends to co-precipitate with Fe as well (Zhang and Sparks, 1990), which might be a reason for the Se removal of 30 % in the wash step. Thus, Fe plays a major role for the success of the purification efficiency. Additional explanations for the high losses in the wash phase might be the incomplete oxidation to selenate leading to reduced sorption affinity as well as the block of limited sorption spaces by chromate and other oxyanions. Both hypotheses are unlikely, because Se species measurements on selected samples showed that Se was completely available as Se(VI) (Appendix IV,

Table IV-12). Furthermore, the detectable removal pathways in the natural plant samples were very similar despite of low As, Cr and Ge contents (Appendix IV, Table IV-6). Sorption space being a limiting factor is improbable, because these were quantified as 1.44 meq for 1.2 mL resin suspension by BioRad (2011). This capacity by far exceeds the amount of potential oxyanions added (Appendix IV, Table IV-6).

Method (B) – thiol retention

Figure 34 shows the removal pathways of the elements using the ICP multi-element standard and method (B).

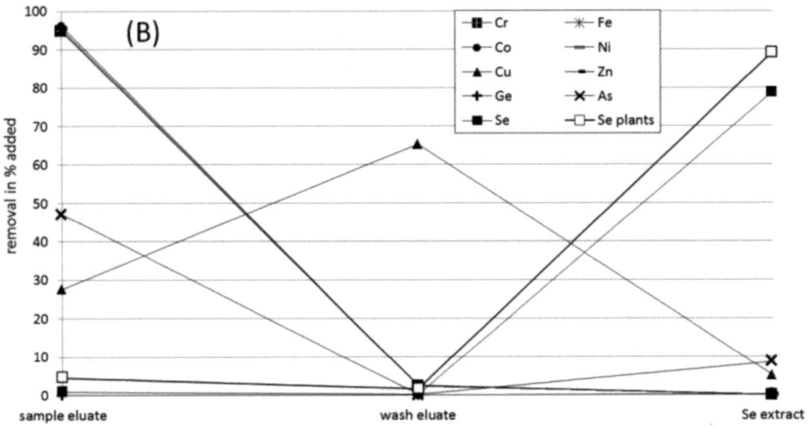

Figure 34: Matrix element and Se ratios determined (related to initial Se in sample) in each step eluate of method (B) derived from the ICP multi-element standard and for Se only for plant digests (raw data available in Appendix IV, Table IV-7).

All elements added were removed by >96 % in the *sample eluate* step, and by the residual few percent in the *wash* step. Exceptions were Cu, As and Ge. Cu was retained in the column by >70 %, but the dominant part (65 %) was removed in the wash step. According to Deratani et al. (1983) and Hultberg et al. (1997) Cu^{2+} has a high affinity to form Cu^{+} complexes with thiol groups by redox reactions. Additionally, Cu is affine to form soluble $(CuCl_4)^{2+}$ complexes (Deratani et al., 1983). Probably, at presence of 1M HCl Cu prefers the formation of thiol-Cu complexes whereas other thiol affine metals (e.g. Cr, Fe, Co) already form soluble chloride complexes. At 6M HCl the chloride is more competitive regarding Cu than thiol groups, which induced the formation of mobile $(CuCl_4)^{2+}$. Arsenic was removed by less than 50 % into the *sample eluate* and by a negligible amount into the *wash eluate*. Residuals of almost 10 % were found in the extraction phase. The major fraction of Ge, apart from some residuals in the extract, was retained in the column even after extraction. Aside from Se, Ge and As were reported to be affine to thiol groups as well and to form similar bonds (Elwaer and Hintelmann, 2008c). Therefore As and Ge might be partly remobilized together with Se during extraction. Arsenic was reported to have a lower thiol affinity than Se, explaining the removal of the main fraction in the sample eluate phase. According to Elwaer and Hintelmann (2008c) the affinity of Ge to the thiol groups was even higher than that of Se, why the dominant fraction probably stayed bound to the TCP. Se could almost be totally retained and extracted by about 80 % (Figure 34) with 1 % Se found in the sample and wash eluates together, indicating that the about 20 % left were still bound to the TCP after extraction. Intensifying the extraction process e.g. by stronger acids, prolonged boiling time or a third extraction process would probably increase the Se recovery, but also mobilize Ge and As from the TCP into the extraction solution and increase the TOC by cellulose residuals. Eichhorn (2014) investigated the purification of As rich soil digests with method (B) and tested an increase of TCP amount to elevate Se recovery. This did not increase Se yields significantly, but led to higher retention of As and, as a consequence, to higher mobilization into the extraction phase.

At presence of 1M HCl, Fe available as Fe^{2+} (Takeno, 2005), tends to form chloride complexes rather than binding to thiol groups or precipitation as oxyhydroxide (Deratani et al., 1983; method (A)). Therefore no remarkable Fe co-precipitation effects could be detected in contrast to method (A).

Method (C) – hydride separation

Figure 35 shows the removal pathways of the elements using the ICP multi-element standard and method (C).

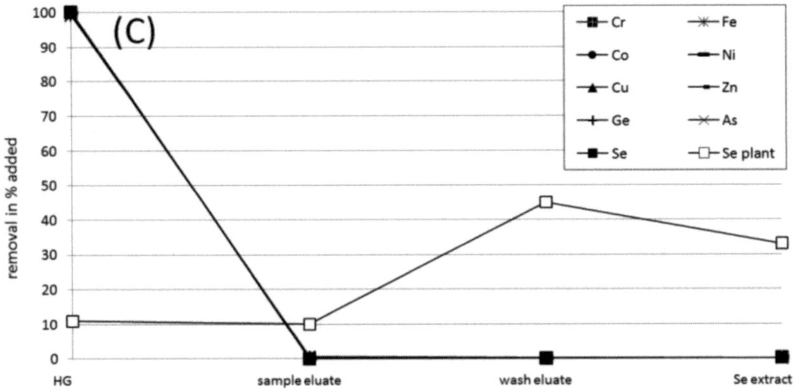

Figure 35: Matrix element and Se ratios determined (related to initial Se in sample) in each step eluate of method (C) derived from the ICP multi-element standard and for Se only for plant digests (raw data available in Appendix IV, Table IV-8).

For the multi-element solution with method (C) all matrix elements added were fully removed in the HG step, including hydride forming elements As, Ge and Se. Welz and Melcher (1984) performed comprehensive tests with hydride inhibiting elements regarding Se. They discovered that HG was very sensitive towards high concentrations of Fe, Co, Ni and Cu, which might suppress it to a minimum. One probable reason for the suppression is catalytic decomposition of the $NaBH_4$, which is essential for HG, in the presence of Co(II), Cu(II), Fe(III) and Ni(II) (Kirkbright and Taddia, 1978). Additionally precipitation of those metals and subsequent capture and decomposition of H_2Se might cause low Se recoveries (Welz and Melcher, 1984). According to this study, the HG rates decrease exponentially from concentrations of 70-3,000 µg L^{-1} for Cu(II), 400-4,000 µg L^{-1} for Ni(II), 5,000-

200,000 µg L^{-1} for Co(II), 3,000-4,000 µg L^{-1} for Fe(II) in the sample matrix. These ranges derive from experiments with 0.5M HCl respectively 5M HCl matrix, meaning that the maximum concentration for the setup in this study with 2M HCl will probably range in the center. However, the multi-element standard exceeded maxima with several of those metals as its concentration was 16,500 µg L^{-1} each. Higher acid (HCl) concentrations in the initial sample could reduce this negative effect, because those metals will stay in solution to a higher extent (Welz and Melcher, 1984). Welz (1983) showed an additional inhibiting effect caused by even low amounts of As(III). At an absolute As content of 1.1 µg As the HG rate drops to 20 % of its initial value with further decreasing tendencies. An amount of 10 µg As – as it was contained in the multi-element standard – could have impeded H_2Se generation as well. Those mechanisms very likely caused the relatively low Se recovery with SGR-1, because it contains amounts of Fe and As that exceed the tolerable amounts by far. Additional inhibiting effects might be caused by HF residuals from digestion: HF was shown to suppress HG as well (Welz, 1983). For Se free plant digests doped with Se standard, on average 89 % of Se was recovered by HG. This fraction was then divided between sample eluate, wash eluate and Se extract during anion exchange (Figure 35). The largest fraction of 45 % was removed in the *wash* step. One probable reason for the limited Se recovery within anion exchange was the incomplete conversion into Se(VI) and therefore a reduced sorption affinity, as Se(IV) tends to adsorb as less stable outerspheric monodentate complex to the resin surface (Zhang and Sparks, 1990). Another possible reason was the partly reduced functionality by the strong alkaline sample matrix. Improvements could be achieved through an additional oxidation step with $K_2S_2O_8$ prior to anion exchange and/or a neutralization of the sample.

Method (C) was shown to be efficient for samples with moderate amounts of matrix elements, but not for samples with high Fe, Co, Ni, Cu or As fractions. For those samples a previous anion exchange step could be helpful as it will decrease metals to a minimum, so that HG will not be inhibited any more. In contrast, this method works very well for plant digests as all matrix elements and residual organic compounds are filtered out (chapter 4.4.6) while keeping Se recovery in an acceptable range.

4.4.5 Se recoveries

At first, the method described by Ellis et al. (2003) was tested and resulted in comparably low and bad reproducible Se recoveries (Table 9) with high matrix residuals (Appendix IV, Table IV-6). As a consequence, optimization approaches were tested with plant digests. The first variation (I) regarded the resin that was poured into the column and settled by gravity in the first place. This might lead to inhomogeneous distribution, preferential flow paths and dead end pores. Therefore a heterogeneous chemical environment within the packing material could have caused the passing through of sample parts without any contact to the resin's surface. In order to mitigate those issues the resin was compressed with a stirring rod after it was poured into the columns (modification (a)) resulting in a moderate success in reproducibility, but lower Se recovery (Table 9). Another limitation of the method by Ellis et al. (2003) was the release of Se from the column in the *wash* step, which was a significant to huge fraction of up to 45 % (Figure 35, Appendix IV, Table IV-6). In a second variation (II) the *wash* step was left out to keep Se in the column and the sample was taken up in 0.1M HCl instead of H_2O to keep matrix elements in solution (modification (b)). This led to higher recoveries, but lower reproducibility (Table 9). In acidic conditions the prevalence of selenate within the Se fraction is more unlikely than in neutral ones, which reduces the sorption affinity (chapter 4.4.1). Fractions of matrix elements in the extraction solutions also grew. Another approach with the same intention was the replacement of 0.1M HCl by H_2O within the wash step in order to avoid accidental remobilization of adsorbed Se by anion exchange with Cl^- (modification (c)). This modification was combined with (a) to Variation III, which is the most successful one regarding Se recovery and reproducibility (Table 9). Furthermore the residual matrix elements in the extraction solution stayed similar to the method by Ellis et al. (2003) and

Variation I. For that reason Variation III was used in method (A) in all further experiments.

Table 9: Se recoveries using original method described by Ellis et al. (2003) as well as three variations (I-III) regarding three modifications: (a) compression of poured resin, (b) leave out wash step, (c) replace 0.1M HCl with H_2O in wash step (-inhomogeneous +homogeneous) (raw data available in Appendix IV, Table IV-9)

Procedure	**Ellis et al. (2003)**	**Variation I**	**Variation II**	**Variation III**
Modification	no	(a)	(b)	(a)+(c)
Flow rate homogeneity	–	+	–	+
Se recovery [%] (n=4)	32.8 ±20.4	26.2 ±13.5	57.4 ±36.2	**66.0 ±8.77**

Table 10 shows the Se recoveries dependent on sample matrix and purification method.

Matrix free solution containing 1 µg Se resulted in recoveries of 100 % within method (A). In contrast, multi-element solution doped with 10 µg Se reached recoveries of only 45 %. Thereby it was shown that the method works quite well in principle, but either suffers from matrix effects or is only suitable for low to moderate Se concentrations. Tests with plant samples digested according to Bell et al. (1992) and containing ~1 µg Se gained Se yields of only 40 % with bad reproducibility. Again, Se free plant digests doped with 1 µg Se standard gained full recovery. Those results imply that the initial Se species is of essential importance for Se retention. Samples rich of natural Se contain a high fraction of organically bound Se. This could be limiting for the validity of the method, as Double Spike must fully equilibrate for mass bias correction (chapter 3.2.7): this cannot be guaranteed for plant samples digested according to Bell et al. (1992). Se_{org} compounds are neglected within this method, confirming the importance of a thorough digestion procedure (chapter 4.3). However, using digestion after Kopp (1999) will increase the probability of valid plant sample treatment as suggested in chapter 4.3. Matrix effects of phytoagar seem to be more significant, because Se free phytoagar extracts doped with 1 µg Se only show average recoveries of 49 % with bad reproducibility. It is hereby assumed that the well-connected long-chain hydrocarbons that are still available in solved form in the extracts might have changed or covered the active sorption surface of the resin. Phytoagar extracts were previously experienced to have adhesive properties (chapter 4.2). Thus, Se species and matrix residuals, especially organic compounds, play an essential role for the efficiency of method (A), which is investigated in detail within chapter 4.4.3. All in all, method (A) can be characterized as principally working with low to moderate Se concentrations, but being highly sensitive on matrix composition.

Within method (B), all sample matrices tested resulted in Se recoveries of 72–89 % with generally good reproducibility. In contrast to method (A), no dependencies on Se concentration and species as well as on matrix elements or organic compounds could be detected. This implies a good reliability and a universal applicability regarding Se recovery and the matrices tested including plant digests and phytoagar filtrates.

For method (C) Se recoveries of HG – without subsequent anion exchange – are regarded in addition to the total Se recovery, because the latter was not detectable for multi-element standard (chapter 4.4.4) and not determined for pure Se solution. Pure Se solutions (1 µg) were recovered by average 82 % with HG, whereas the value was even significantly higher und good reproducible with Se doped plant digests. In this case the digest HNO_3 matrix might have enhanced the hydride formation rates as reported by Welz and Melcher (1981). Unfortunately method (C) was not tested systematically with natural Se rich plant samples, but it can be assumed that the Se species plays a major role here as well. Fitzpatrick et al. (2009) stated that the HG reaction (Equation (11)) exclusively occurs with Se(IV). Nevertheless, plants digested after Kopp (1999) will meet this prerequisite with high probability. Se recoveries with Se doped phytoagar extracts were on a slightly higher level than plants after HG, on average 92 %. HNO_3, added to phytoagar filtrates at the organic destruction step, might have had an enhancing effect in analogy to plant digests. After anion exchange, the Se recoveries of plant and phytoagar samples were on a similar level and additionally slightly higher than this of method (A) applied to Punjab plants.

This might be caused by Se being totally available in inorganic form in the Se doped plant and phytoagar samples, while being available as Se_{org} by average 8 % in Punjab plants (chapter 4.3). Using multi-element standard, Se yields were close to 0 %, and with SGR-1 they were at relatively low level as well. As already explained in chapter 4.4.4 high concentrations of diverse metals were probably responsible for high Se losses within HG (Welz (1983); Welz and Melcher (1984)).

Table 10: Se recoveries and external reproducibility tested with purification methods (A), (B) and (C) depending on sample matrices (*ICP multi-element standard containing 100 µg Na, Mg, Al, Ca, Cr, Fe, Co, Ni, Cu, Zn and 10 µg As, Ge, Se, all other samples initially contained 1 µg Se; ** Se recovery after HG *** Se recovery after HG and anion exchange) (raw data available in Appendix IV, Table IV-9)

method	**(A) – anion exchange**		**(B) – thiol retention**		**(C) – hydride separation**		
sample matrix	**Se recovery [%]**	**n**	**Se recovery [%]**	**n**	**Se recovery HG** [%]**	**Se recovery total*** [%]**	**n**
pure Se (Roth)	100 ±0.8	3	72 ±1.1	3	82 ±5.6	n/a	2
MS* (Roth)	45 ±7.0	8	79 ±13.9	7	0.05 ±0.01	n/a	2
Se rich plants (Punjab)	40 ±19.6	9	89 ±18.8	7	n/a		
Se free plants + doped Se	99 ±1.9	3	74 ±1.0	3	87 ±7.3	53 ±20.1	10
Phytoagar filtrate	49 ±16.9	3	77 ±0.7	3	92 ±2.9	54 ±2.3	2
shale (SGR-1)	n/a		n/a		n/a	27	1

4.4.6 Residues in purified samples

Table 11 shows the matrix residuals in the extraction samples of method (A) and (B) as well as all sample matrices used.

Table 11: Absolute contents of residuals [ng] in purified extraction sample using methods (A), (B) and (C) and ICP multi-element standard (MS), plant digest from the Punjab plants (pp), phytoagar (p) as well as plants (cp) from the cultivation experiments as matrices; for (C) no experiments with Punjab plants were performed, but with SGR-1 green river shale (USGS reference material). The tolerance test was performed to evaluate purification sufficiency for analytical purposes (chapter 4.5.1) (raw data available in Appendix IV, Tables IV-6 to IV-8) (Standard – not exceeding tolerance test; Cursive – slightly exceeding tolerance test (<200 %); Bold – significantly exceeding tolerance test (>200 %)); *Fe exceeding caused by incidentally high blank, phytoagar extract only contained small traces

[ng]	**n**	**Cr**	**Fe**	**Co**	**Ni**	**Cu**	**Zn**	**Ge**	**As**
(A)$_{MS}$	9	**10640** ±1590	**223** ±483	<0.02	*4.8 ±1.7*	*18.5 ±14.7*	12.4 ±3.9	<0.1	**44.9** ±27.8
(A)$_{pp}$	9	*1.9 ±2.3*	30.7 ±18.4	<0.02	1.7 ±1.1	2.6 ±1.8	58.3 ±41.3	<0.1	<0.1
(A)$_{p}$	18	1.1 ±0.8	29.2 ±26.5	<0.02	0.1 ±0.9	0.9 ±2.2	10.5 ±18.3	<0.1	0.6 ±0.5
(A)$_{cp}$	9	0.5 ±0.5	45.8 ±35.2	<0.2	0.0 ±0.1	3.1 ±2.8	124 ±180	<0.1	0.4 ±0.4
(B)$_{MS}$	9	**7.8 ±7.2**	47.9 ±26.4	*1.31 ± 0.50*	*5.7 ±3.6*	**5280** ±326	81.7 ±43.6	**79.2** ±73.9	**2750** ±2360
(B)$_{pp}$	9	0.8 ±2.3	4.1 ±40.8	<0.02	1.1 ±0.9	0.5 ±1.1	248 ±248	<0.1	0.7 ±0.8
(B)$_{p}$	18	*2.5 ±2.9*	**137* ±227**	<0.02	1.6 ±1.0	*24.5 ±36.9*	34.5 ±193	<0.1	5.4 ±1.1
(B)$_{cp}$	9	0.3 ±1.3	11.7 ±12.7	<0.02	0.2 ±0.6	1.2 ±1.0	37.8 ±57.7	<0.1	4.3 ±1.0
(C)$_{MS}$	2	**34.7** ±1.2	6.13 ±1.41	0.41 ±0.20	**59.0** ±1.8	<0.1	6.2 ±0.7	<0.1	2.8 ±0.0
(C)$_{p}$	2	**53.8** ±10.0	15.5 ±5.32	0.53 ± 0.00	**112** ±57.8	0.5 ±0.3	11.1 ±0.2	<0.1	<0.1
(C)$_{cp}$	2	**35.8** ±1.6	32.1 ±11.8	*0.71 ± 0.31*	**174** ±48.3	13.2 ±8.4	9.0 ±1.7	0.4 ±0.0	<0.1
(C)$_{SGR-1}$	2	**25.5** ±6.6	20.7 ±2.90	0.46 ± 0.10	**178** ±11.7	5.2 ±5.2	27.2 ±17.3	0.3 ±0.0	1.6 ±0.3
tolerance test	**1**	**1.39**	**52.3**	**0.57**	**3.1**	**13.3**	**686**	**0.1**	**4.9**

Matrix residuals are generally low regarding the initial matrix element concentrations as well as the laboratory blanks. Only few elements of particular samples significantly exceed the tolerance test, which was performed to estimate the analytical sufficiency of matrix removal (chapter 4.5.1) (Table 11). This applies for Cr, Fe, Cu, Ge and As in the multi-element standard solution, which was over-concentrated anyway to examine removal pathways and therefore did not represent natural samples. Significant amounts of those elements are usually not available in plants and phytoagar as shown by residual concentrations in those, which are mainly caused by blanks in the open laboratory (chapter 4.1). Method (B) generally results in higher matrix element residuals than (A) and (C) due to the boiling extraction (chapter 4.4.1), which probably caused mobilization of impurities from the cellulose powder. However, any method supplies sufficient purification for Se isotope analytics as they generally do not exceed the tolerance test and will in no case reach critical concentrations for the on-line HG. Potential isobaric interferences such as Ge are on a very low level for any sample except the over-concentrated multi-element standard purified with method (B). Nevertheless matrix residuals must be continuously monitored. Especially samples with elevated Cr, Fe, Cu, Ge or As must be regarded on those residuals.

Table 12 shows the TOC residuals in the Se extracts samples with methods (A), (B) and (C).

Table 12: Residual TOC concentrations in plant and phytoagar samples purified with method (A), (B) and (C) (raw data available in Appendix IV – Table IV-10)

method	**(A) – anion exchange**		**(B) – thiol cellulose powder**		**(C) – hydride separation**	
matrix	**TOC [mg L^{-1}]**	**n**	**TOC [mg L^{-1}]**	**n**	**TOC [mg L^{-1}]**	**n**
plants	7.8 ±1.8	3	19.9 ±4.7	3	<0.9	3
phytoagar	73.0 ±32.0	3	49.9 ±25.4	3	<0.9	3

For samples highly enriched in organic compounds the TOC of the purified samples was measured in order to evaluate the removal efficiency of the methods and the residual analytical disturbance potential of the samples. The samples derive from cultivation experiments and therefore represent samples measured on Se isotope composition. TOC residuals in plants are generally low, but measurable within method (A) and (B) and on average still 14 % respectively 35 % of the digest's average TOC. The chromatographic methods offer pore space in which organic residuals might have been retained as well as surfaces to which they might have been adsorbed during purification. Afterwards they might have been extracted together with Se. The higher value for method (B) might be caused by cellulose from TCP mobilized during the extraction phase. Phytoagar extracts released in absolute numbers 5 to 10 times more TOC into the purified sample. Related to TOC in the initial phytoagar extracts, those fractions of 15-22 % were comparable to plant digests. The value for method (A) was significantly higher, which might be caused by higher reaction surface where the hydrocarbons might have stuck to or were retained in. Another aspect might be the additional organic destruction with HNO_3 boiling as extraction step in method (B) (chapter 4.4.1). TOC residuals in phytoagar could be reduced by an extended HNO_3-H_2O_2 organic destruction step prior to purification. In samples purified with method (C) residual TOC could be detected neither in the plant digests nor in the phytoagar extracts.

4.4.7 Method evaluation

In principle, all methods are suitable to purify samples for Se isotope analytics regarding the criteria set in advance, namely Se recovery, matrix element residuals and TOC residuals. Particularly for methods (A) and (C) Se recovery was shown to be highly dependent on Se species availability, whereas method (B) yielded similar high values for any method tested. In most samples, Se recoveries were sufficiently high to measure Se isotope ratios as the Double Spike will correct mass bias induced by losses (chapter 3.2.7). However, for limited sample amounts or Se contents, the method should be chosen addressing this criterion as well. Method (B) emitted higher impurities into the purified samples than (A) and (C), but all values were far below critical concentrations. Each method was characterized by particular elements that might become critical if available in certain concentrations. These are Cr for method (A), Ge, As and Cu for method (B) as well as transition metals (Fe, Co, Ni, Cu) and hydride generating elements (As, Ge) for method (C). In the open laboratory certain ubiquitary elements were slightly elevated compared to clean laboratories, e.g. Fe and Zn, but they should not cause analytical problems in the ranges measured (chapter 4.5). Organic residuals were found in significant amounts in purified samples from (A) and (B), but below the detection limit in samples from (C). Validation tests (chapter 4.5.3) will eventually prove if matrix element and TOC removal is sufficient for the respective method.

4.5 Analytical quality control

The actual suitability of the preparation (chapters 4.1 and 4.2) and purification methods (chapter 4.4) for accurate, precise and valid Se isotope ratios was assessed by analytical quality monitoring on MC-ICP-MS. Three parameters were checked independently: the presence of measurable isobaric interferences, the internal and external reproducibility as well as the validity of the Se isotope ratios measured.

4.5.1 Isobaric interferences

Matrix compounds become critical for Se isotope analytics, if they form interferences on masses that were used for Se isotope detection or monitoring (72, 73, 74, 77, 78, 80, 81, 82 and 83, chapter 3.2.3). With the exception of Ge, potentially critical compounds are mass interfering only as hydride or oxide (Table 2). Due to the analytical setup (chapters 3.5.1 and 3.5.5), hydride formation in the plasma is suppressed to a minimum. The on-line HG system filters out non-hydride generating sample compounds and therefore reduces the potential of metals to form mass interfering oxides. However, to exclude those influences and to test if samples were sufficiently purified, the elemental masses of each potentially critical hydride or oxide molecules were tested on signal intensity. Therefore masses 56 ($^{56}Fe^{16}O$), 57 ($^{57}Fe^{16}O$), 58 ($^{58}Fe^{16}O$; $^{58}Ni^{16}O$), 59 ($^{59}Co^{16}O$), 60 ($^{60}Ni^{16}O$), 62 ($^{62}Ni^{16}O$), 64 ($^{64}Ni^{16}O$; $^{64}Zn^{16}O$), 65 ($^{65}Cu^{16}O$), 66 ($^{66}Zn^{16}O$), 67 ($^{67}Zn^{16}O$) and 75 ($^{75}As^{1}H$) were determined. Additionally those masses, which tend to inhibit HG, but have no direct interference potential (Cr, Mn and Cu oxides), were measured: 50, 52, 53, 54, 55 and 63. Purified plant samples with highest residual concentrations regarding critical compounds were used in order to exclude interfering impacts on any sample type (Table 11). In the cultivated plant samples no Ge was expected. However, due to efficient mathematical Ge corrections (chapter 3.2.8.1) relatively high amounts are tolerated.

For none of the masses tested significant signal-to-noise ratios compared to NIST3149 and on-peak-zeros could be observed. This reveals a certain tolerance for matrix residuals that were contained in the samples tested due to purification residuals and open laboratory blanks (Tables 4 and 5). The analytical setup suppresses critical residuals and therefore allows preparation in open laboratories and incomplete matrix removals. The actual extent of tolerance was not investigated, but it proved to be sufficient for plant and phytoagar matrices regarding all purification methods tested as the highest matrix element concentrations available did not indicate any limitations.

4.5.2 Reproducibility

Internal reproducibility means in this case the analytical ability to replicate an isotope ratio using aliquots of the same prepared sample in the same or different measurement runs. External reproducibility describes the ability to gain identical isotope results from the same sample, prepared and purified independently. Both tests were performed with phytoagar and plant material. For samples purified with method (B) the internal reproducibility was determined independently for two plant samples (0.050 ‰ and 0.024 ‰) and two phytoagar samples (0.013 ‰, 0.004 ‰). Samples prepared with method

(C) showed internal reproducibilities in a similar range (plants: 0.040 ‰; phytoagar 0.051 ‰). Stating a NIST3149 reproducibility of 0.2 ‰ (chapter 3.2.1), all samples are clearly within the standard internal reproducibility, meaning that no matrix impacts on internal reproducibility could be detected. This applies for all samples measured, presented in chapters 4.5.3 and 5.6 (raw data in Appendix IV, Table IV-11).
External reproducibility was tested using Se free plant and phytoagar matrices independently doped with NIST3149, then spiked, digested/extracted afterwards and finally purified with (A), (B) and (C) (chapter 4.5.3). Results are given in Table 13. External reproducibility with methods (A) and (B) applied are on a high level and by far exceed the NIST3149 reproducibility of 0.2 ‰. (C) purified samples showed a much better external reproducibility, in which only plant samples spiked before digestion exceeded that limit. In any case, plant samples spiked before digestion had significantly higher external reproducibility than those spiked afterwards. Obviously small differences in Double Spike losses during digestion have exceptionally high impacts on the external reproducibility. In plant material spiked after digestion an external reproducibility of 0.2 ‰ was detected using eight independently prepared samples. This value being within the range of NIST3149 regarding a relatively high number of reproductions confirms the reliability of the treatment and purification procedures concerning reproducibility. This applies for phytoagar as well, which external reproducibility was significantly lower with 0.1 ‰, based on a lower number of reproductions as well.

4.5.3 Validity

Method validation is usually performed with certified reference materials having a comparable matrix to the target samples. For plant and phytoagar matrices, no reference material is certified on Se isotope ratios or even measured on that by external laboratories. For this reason an alternative validation test approximating reference conditions was set up. This test should include the target matrices as well as Se in a known and certified isotope composition. Therefore Se free plant tissue and phytoagar samples were taken from the cultivation experiments (blank boxes, chapter 5.1) and doped with NIST3149 before and after plant digestion (in parallel setups) and directly after phytoagar vacuum filtration. After letting NIST3149- and Double Spike-Se equilibrate for 24 hours in closed beakers at room temperature, samples were purified according to chapters 4.3.2.1 to 4.3.2.3 and measured on Se isotope composition afterwards. As NIST3149 was also used as standard reference for δ^{82}Se calculations, the target δ^{82}Se to prove validity was 0 ‰. Figure 36 illustrates the validation test procedure. Table 13 shows δ^{82}Se values measured in the prepared plant and phytoagar samples dependent on the purification method as well as their external reproducibility.

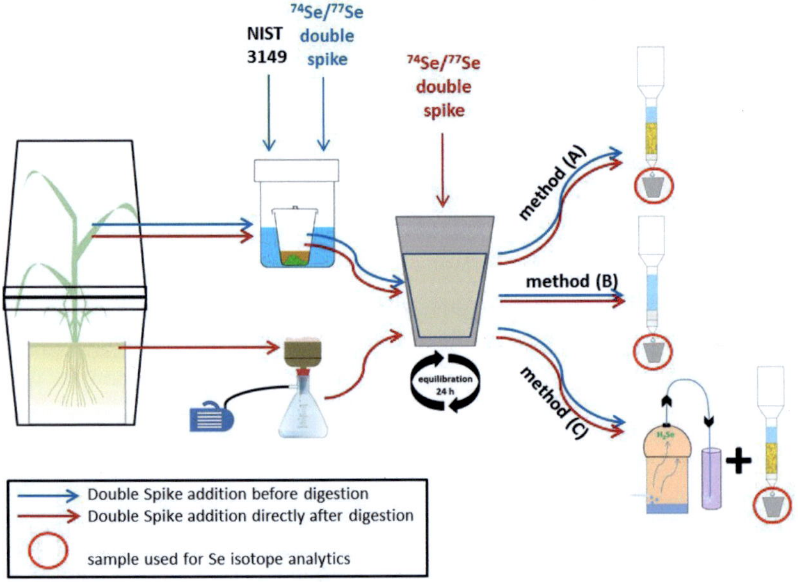

Figure 36: Process scheme of validity tests in dependence on matrix, sample treatment procedure, purification method and date of Double Spike addition.

Table 13: $\delta^{82}Se$ and external reproducibility of validation test samples dependent on purification method, sample matrix and date of Double Spike addition (*Double Spike added before digestion, elsewhere Double Spike added after digestion) (raw data and internal errors available in Appendix IV, Table IV-11)

Sample matrix	Purification method	$\delta^{82}Se$ [‰] (average)	External reproducibility	Repetitions (n)
plant	(A)	2.0	1.6	2
phytoagar	(A)	5.5	1.2	2
plant*	(B)	11.6	10.5	6
plant	(B)	25.5	5.7	6
phytoagar	(B)	28.4	7.9	2
plant*	(C)	0.6	0.7	2
plant	(C)	0.2	0.2	8
phytoagar	(C)	1.1	0.1	2

Average $\delta^{82}Se$ values measured in samples purified with (A) and (B) highly differed from the target value 0 ‰ for both plant and phytoagar, whereby (B) samples were even more deviating and worse reproducible than (A) samples. In contrast, (C) derived samples showed values between -0.2 ‰ and +2.1 ‰. Thereby plants yielded only slightly above 0 ‰ and phytoagar samples were significantly above the target value with average +1.1 ‰. In single cases, plant samples spiked before digestion had a $\delta^{82}Se$ value closer to 0 ‰ than the ones spiked afterwards, but the majority had a higher distance to 0 ‰, and the external reproducibility was significantly higher as well (Table 13, Appendix IV, Table IV-11).

Figure 37 shows the $\delta^{82}Se$ values of the individual samples, NIST3149 and MH495 standards as well as their correlation with the instrumental fractionation factor β_{instr} (chapter 3.2.8.2, Equation (50)). The correlation between those two parameters is obvious for (A) and (B) samples, but not significant for (C) samples as well as for NIST3149 and MH495 matrix free standard. This indicates that matrix effects are responsible for the correlation and, probably related to that, for the invalid $\delta^{82}Se$ values.

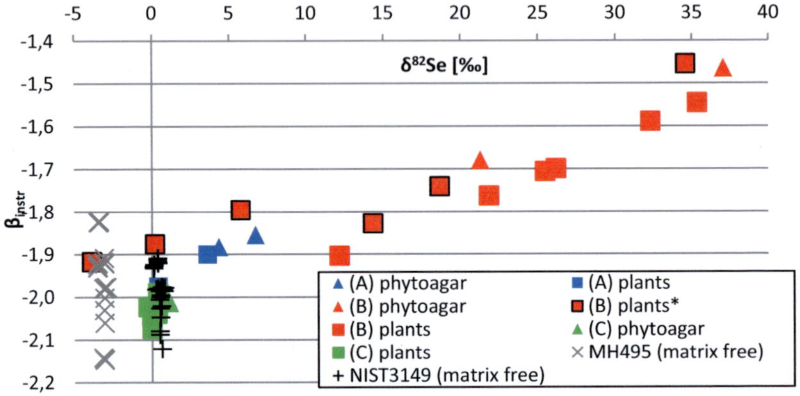

Figure 37: Individual δ^{82}Se values in dependence on the instrumental mass bias factor β_{instr} for validation tests processed with purification methods (A), (B) and (C) and both plant and phytoagar matrices. For comparison, NIST3149 and MH495 matrix free Se isotope standards are added. The vertical line represents the target value 0 ‰ for the validation tests. The Double Spike was added after digestion/extraction regarding all samples except the ones marked with * in which it was added before digestion (raw data available in Appendix IV, Table IV-11).

The correlation between β_{instr} and δ^{82}Se for (A) and (B) samples along with the invalid results indicate that the instrumental mass bias correction, performed by the Double Spike, did not work properly. The calculation of β_{instr} is based on signals measured on masses 74 and 77 (chapter 3.2.8.2, Equation (50)), which could have been interfered during measurements. Furthermore the calculation of δ^{82}Se could lead to incorrect results by interferences on all Se and monitor masses concerned (chapter 3.2.8). Significant amounts of TOC were measured in the (A) and (B) purified samples (Table 12). Organic compounds have a high potential to interfere the HG process and even form hydrides themselves (e.g. decalin, methylcyclohexane) (Karadjova et al., 2006; Saito et al., 2008). The gaseous compounds formed might then be transported via the plasma into the detector and formed isobaric interferences on Se masses. This hypothesis is confirmed by the results of method (C) derived samples that did neither contain detectable amounts of TOC nor show dependencies of δ^{82}Se and β_{instr} or highly invalid results. Furthermore (B) tends to have higher TOC residuals and higher δ^{82}Se deviations from 0 ‰ although there is no clear correlation between residual TOC and δ^{82}Se or β_{instr}. Another reason for invalid δ^{82}Se values could be the inhibition of full equilibration between sample- and Double Spike-Se by mass dependent interaction of Se with organic matter. Se tends to interact with organic compounds to differing extent and binding mechanism, which is dependent on the Se species (Wasilewska et al., 2002; Zsolnay, 2003). Differing addition date of sample-Se – in this case NIST3149 – and Double Spike-Se could have let to differing interaction mechanisms and therefore incomplete equilibration though they were likely available in the same Se species Se(IV) (chapter 3.2.7). As a consequence, instrumental mass bias and artificial Se isotope fractionation during sample preparation could not be corrected properly. This hypothesis is confirmed by varying dependencies of δ^{82}Se and Se recovery regarding the preparation step. Se recovery from HG only somehow correlates with the deviation of δ^{82}Se from 0 ‰, whereas the Se recovery of the anion exchange performed afterwards did not seem to have any influence on the precision (Figure 38).

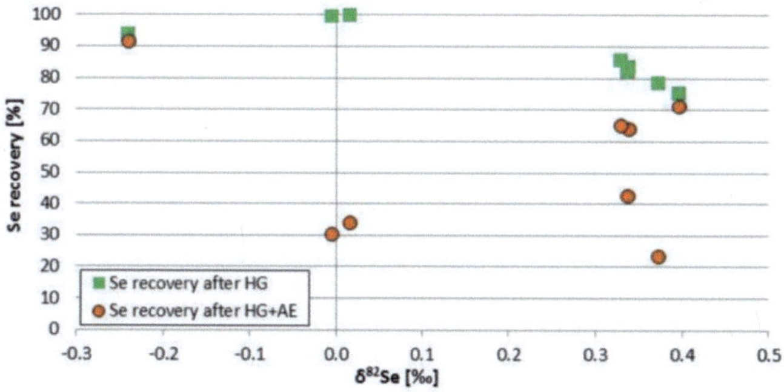

Figure 38: Individual δ^{82}Se values dependent on Se recoveries after HG and total Se recovery after HG and the subsequent anion exchange (AE) step in plant samples purified with method (C) (raw data available in Appendix IV, Table IV-11).

However, the deviation of δ^{82}Se from 0 ‰ shows a clear matrix dependency within method (C) (Table 13), indicating plant and phytoagar specific interferences or interactions. Invalid results are probably caused by both organic isobaric interferences and incomplete Double Spike-/sample-Se equilibration. Their extent likely depends on the amount of TOC and the particular matrix structure. However, no direct correlation between TOC and δ^{82}Se could be detected as only single samples were tested on their TOC residuals. Completing this data basis could provide insight into the role of TOC in Se isotope analytics. To find out about the particular role of organic isobaric interferences and to increase the precision, high resolution MC-ICP-MS measurements might be an approach.

However, the entire sample preparation procedure including plant digestion after Kopp (1999), phytoagar vacuum filtration as well as purification according to method (C) turned out to be suitable for Se isotope analytics. It facilitates the mass bias correction mechanism and produces valid results with good external reproducibility and a precision of 0.2 (±0.2) ‰ for plants. This precision is probably sufficient to detect Se isotope variations in plants, as Herbel et al. (2002) and Schilling et al. (2015) reported δ^{82}Se differences in soil water and plant tissue of -1.1 ‰ respectively +2.4 and +3.2 ‰. Plant internal Se isotope variations have not been published yet. The precision of phytoagar should be increased for reasonable Se isotope determinations. As method (C) purification worked with a comparable efficiency to plant matrices, TOC might be further reduced after vacuum filtration to gain more precise data (chapter 4.4).

To validate the preferential purification method (C) with natural Se rich samples, SGR-1 reference material (USGS) was used. This material was not certified on Se isotope composition, but previous studies measured it with differing analytical setups (Table 14) and therefore provided a certain data basis that does not exist for any plant reference. Figure 14 and 15 provide information on the analytical methods used in these studies.

Table 14: $\delta^{82}Se$ values for SGR-1 standard reference material (USGS) reported by several studies and their external reproducibility (*conversion from MERCK S.A. Tritrisol (element standard solution) to NIST3149 scale after Carignan and Wen (2007), ** conversion from MERCK S.A. Tritrisol (element standard solution) to NIST3149 scale after Layton-Matthews et al. (2006), *** calculated from reported value for $\delta^{82/78}Se$ according to Stüeken et al. (2013))

study	**$\delta^{82/76}Se$ [‰]**	**external reproducibility**	**n**
Rouxel et al. (2002)*	+0.54	0.37	not reported
Layton-Matthews et al. (2006)**	+0.62	0.32	3
Schilling et al. (2011a)	+0.2	0.1	4
Mitchell et al. (2012)	-0.2	0.05	11
Stüeken et al. (2013)***	-0.09	0.28	26
Pogge von Strandmann et al. (2014)	+0.25	0.17	16

$\delta^{82}Se$ for SGR-1 ranges closely around 0 ‰ with a slight tendency to positive values and an overall range regarding study internal averages of -0.2 ‰ to +0.62 ‰. This range as well as the external reproducibility within the individual studies is relatively high, which is probably caused by SGR-1 sample heterogeneities, but also differences in sample preparation, analytical setups and correction mechanisms. Although the methods presented in this study are adapted to plant and phytoagar samples, the preparation and measurement of SGR-1 is reasonable to examine the external reliability with regard to organic rich samples. For this purpose 200 mg of SGR-1 was digested with HF-$HClO_4$-HNO_3 on a hotplate as described in chapter 4.4.2. Afterwards it was purified according to method (C) (chapter 4.4.1) and measured on Se isotope composition.

To create an internal Se isotope standard and to provide certified reference plant tissue for this type of analytics, the Se isotope composition of NISTSRM1567a (Wheat Flour, NIST) was additionally determined in this study. This reference was chosen for its reasonably high Se content as well as its high ratio of organically bound Se, making the validation applicable for all Se species supplied (chapter 4.3). 100 mg of NISTSRM1567a was digested in analogy to the cultivated plant samples (chapter 4.3), purified according to method (C) (chapter 4.4.1) and measured within this study. Table 15 contains the $\delta^{82}Se$ values and the external reproducibility for SGR-1 and NISTSRM1567a reference materials.

Table 15: $\delta^{82}Se$ values measured in SGR-1 and NISTSRM1567a reference materials, their external reproducibility and for NISTSRM1567a TOC and Se recovery after HG used as continuous quality monitoring parameter (raw data available in Appendix IV, Table IV-11)

Sample	**digestion**	**$\delta^{82}Se$ [‰]**	**n**	**TOC [mg L^{-1}]**	**Se recovery after HG [%]**
SGR-1	$HF-HClO_4-HNO_3$	0.74	1	n/a	n/a
NISTSRM1567a	acc. Kopp (1999)	0.27 ±0.08	2	<0.9	97.8 ±1.3

For SGR-1 only a single $\delta^{82}Se$ value was available, but this one is within the error range of the studies previously published (Table 14). Therefore it is probable that purification method (C) works for SGR-1, although the Se recovery was relatively low (chapter 4.4.5). NISTSRM1567a showed a very good external reproducibility being within the analytical precision for NIST3149. Analytical quality monitoring parameters made up in chapter 4.5.3 indicated high precision and validity due to no detectable TOC and high Se recovery after HG. Therefore the average $\delta^{82}Se$ determined is reliable and suitable as internal reference and for further Se isotope studies on plants.

5 Determining the relation between Se isotope signatures and metabolic processes in plants

Building on the solid and reliable methodical basis developed and described in chapter 4, the next task was to figure out the relationship between Se metabolic pathways in plants and Se isotope signatures left behind in the plant tissue and growth media. Numerous previous studies (chapters 2.1.4 and 2.2) record the crucial role of pH value and redox potential in soils for the Se transport pathways in the Critical Zone and the uptake and accumulation in different plant parts. The primary reason for their high influence is their determination of Se species availability and transformation. Se species underlie different patterns regarding uptake, transport, accumulation and volatilization into and within plants (chapter 2.2). To differentiate between those processes that in nature occur simultaneously, a *Minimum Parameter* setup was implemented based on a modified concept of Nothstein (2015). Transparent closed boxes were filled with phytoagar that serves as nutrient free growth medium to exclude the influence of soil components (sorption, binding to organic matter) and soil solution (competing ions, solved organic molecules). The phytoagar was doped with Se in varying species (SeO_4^{2-}, SeO_3^{2-}, SeMet), chosen according to their abundance within the Critical Zone and their relevance for plant uptake (chapters 2.1.4 and 2.2). The species each were added in varying concentrations (100, 500, 1000 µg L^{-1}) regarding the optimum uptake rates derived from pre-studies covering a wide concentration range (0-2500 µg L^{-1}) (Nothstein, 2015). Sterilized seeds were planted under sterile conditions to avoid microbial influence. The boxes were airtightly closed afterwards and placed into a climate chamber. This concept was defined for short cultivation periods of total 16 days because of water, CO_2, nutrient and space limitations. The first days and weeks were shown to be critical for Se uptake (e.g. Li et al., 2008; Nothstein, 2015), that is why this time period was sufficient and enabled repetitions in a reasonable time frame. Seeds of *Oryza sativa* were used, which is the predominant sort of rice applied in agricultural systems. Rice plays an exceptionally important role in global nutrition as it is the staple food for more than 50 % of the world's population and provided on average 19 % of nutritive energy and 13 % of nutritive protein (CGIAR, 2013). Rice plants are reliable cultivates, even at minimum parameter conditions and without the supply of external nutrients. Furthermore *Oryza sativa* is a model plant. A huge amount of background knowledge on metabolic mechanisms of macro- (e.g. Yang et al., 2014) and micronutrients (e.g. Arnold et al., 2015) as well as toxins (e.g. Pan-pan et al., 2015) and physical stress factors (e.g. Glaubitz et al., 2015) has already been generated. The Se metabolism within *Oryza sativa* was investigated by several studies as well, e.g. Terry et al. (1992), Zhao et al. (2010), Zhang et al. (2014) and extensively by Nothstein (2015). As mentioned in chapter 2.2.4, Terry et al. (1992) revealed the high quantity of Se volatilization by *Oryza sativa* compared to a variety of other crop species. As volatilization plays an exceptionally high role in the biospheric Se cycle (Lin et al., 2002), a plant with sufficiently high volatilization rate is advantageous to detect relevant process characteristics. However, there are still many knowledge gaps concerning plants in general and rice in particular (chapter 2.2). Arnold et al. (2015) performed zinc and iron isotope studies with *Oryza sativa* cultivations grown under different redox conditions, which revealed varying isotopic signatures among the plant parts and thereby indicated different translocation and grain load mechanism of zinc and iron. Reasonable and valuable results of other non-traditional stable isotope system applications indicate the suitability of *Oryza sativa* for Se isotope studies as well.

According to chapters 2.2.2 and 2.2.3 there are characteristic Se transformation pathways for each Se source species. Figure 15 shows the simplified metabolic pathways that potentially occur. The individual paths underlie quantitative shifts depending on Se species distribution and Se concentration supplied as well as plant species. Based on that, potential pathway schemes for the box setup dependent on Se source species are illustrated in

Figure 39. Transformation processes are categorized with regard to their role for distribution patterns and their relevance for Se isotope fractionation (chapters 2.3.2 and 2.3.3). Thereby, reduction is the key process for isotope fractionation and volatilization retains isotope variations in plant tissue by selectively removing parts of the most reduced species Se(-II).

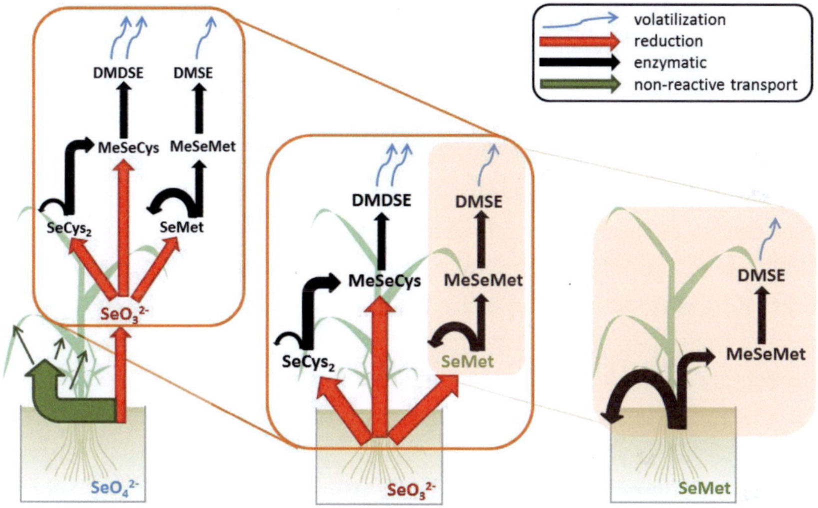

Figure 39: Hypothetical transport pathways of Se taken up by plants dependent on Se species available in the growth medium (qualitative image, arrows not to scale) (data from chapter 2.2, Figure 15).

The prevalence and dominance of particular pathways depend on plant specific metabolisms reacting on Se availability and plant internal Se species distribution (chapter 2.2.3). Knowledge on these could therefore offer an insight into the history of Se accumulation within the plants including Se species availability, concentrations and environmental conditions. Predictions on Se behavior as well as solution to Se related challenges could be developed based on this knowledge. Plant-internal processes can hardly be traced by variances in concentrations or species alone as those parameters are only able to represent one moment in a dynamic system and are in practice difficult to map and hardly representative in single plants. In contrast, Se isotope signatures enable the differentiation between isotope fractionating and non-fractionating processes, whereby a mass balance model provides quantitative data on Se transport by uptake, translocation and volatilization. Previously published data on Se species distribution in plants illustrate the high potentials of Se isotope signatures in tracing Se pathways (chapter 2.2).

According to Figure 11, supplied selenate tends to remain in this species within the plant to 60-80 %. The fraction left is majorly available as organic Se(-II) species reduced in three steps, Se(VI)->Se(IV)->Se(0)->Se(-II). Due to Se isotope fractionation in reduction processes and the enrichment of light isotopes in their products (chapter 2.3.2), the reduced fraction will very likely be isotopically lighter than the selenate fraction. Reduction and organic transformation of selenate is tendentially higher in the roots, while the residual selenate is transported to the shoots. That is why the roots will expectedly be depleted in ^{82}Se compared to the shoots and the Se source. Volatilization will increase this difference as it takes place dominantly via the shoots within *Oryza sativa* (Terry et al., 1992) and exclusively concerns organic Se compounds (chapter 2.2.4). If volatilization plays a significant role, this might be detectable by enrichment in ^{82}Se in the plant compared to the Se source. Thereby the root-shoot difference in δ^{82}Se will directly depend on the volatilization rate as well. As rising Se source concentrations might induce the volatilization of a higher fraction of the organic Se in order to overcome critical Se contents (Figure 39), higher concentrated setups might further enrich heavy Se in the shoots. In this case, a potentially rising importance and quantity of volatilization via shoots might mitigate the root-shoot difference, but increase the source Se-plant difference in δ^{82}Se.

Supplied selenite will be majorly reduced after entering the plant (Figure 11), that is why plant-internal distribution will probably have no detectable Se isotope effect. After reduction, there are three main pathways for organic transformation with differing volatilization tendencies (Figure 39). They will probably shift with Se concentration supplied as well. In the sele-

nite setups volatilization will determine Se isotope variations. As the major fraction remains in the roots after being taken up (Figure 11), volatilization via roots might play a significant role. Thereby, volatilization pathways depending on the plant part might be traceable. As Se will likely be available as a mixture of Se species only fully reduced Se (Se(-II)) will volatilize, the isotopically lighter fraction within the plant part will decrease by volatile emissions. Therefore higher volatilization rates might lead to enrichment in heavier isotopes regarding total Se.
SeMet is already available as fully reduced species that only underlies enzymatic transformations and to some extent volatilization. This setup is applicable to examine if those processes themselves might induce isotope fractionation or if reduction is the only or very major process as assumed e.g. by Schilling et al. (2011b). It furthermore represents the SeMet incorporation as well as the SeMet -> MeSeMet -> DMSe pathway which is inherent in selenate and selenite setups as well and therein applies only for a certain fraction (Figure 39).

5.1 Minimum Parameter Experiments (MinPaX)

The general concept of the cultivation experiments derives from Nothstein (2015) differing in Se concentrations and species supplied. 8.8 g phytoagar powder (Duchefa) was added to 2.2 L H_2O (millipore) to produce 0.4 % semi-solid phytoagar growth medium. This mixture as well as 10 glass beakers (200 mL), 3 glass scaling beakers (100 mL), 3 stirring rods, 10 cleaned Magentaboxes (SigmaAldrich) (chapter 4.1) and a 500 mL bottle with H_2O (millipore) were autoclaved at 120°C for 3 h for full sterilization. Magentaboxes each consisted of two equal parts (6x6x10 cm) combined with a coupler (Figure 41 (c)). After autoclaving, the hot (70°C) and liquid phytoagar was transferred into the glass beakers in 10 200 mL portions under a sterile bench. Sterile filtrated sodium selenate (Na_2SeO_4) standard (100 mg L^{-1} Se) was added to beaker 1 (200 µL), 2 (1000 µL) and 3 (2000 µL), sodium selenite (Na_2SeO_3) standard (100 mg L^{-1} Se) to beaker 4 (200 µL), 5 (1000 µL] and

6 (2000 µL) and SeMet standard (100 mg L^{-1} Se) was added to beaker 7 (200 µL), 8 (1000 µL) and 9 (2000 µL) and thoroughly mixed with sterilized stirring rods. Thereby phytoagar media with the target Se source concentrations of 100, 500 and 1000 µg L^{-1} were made for each species (Figure 41 (a)). Beaker 10 was left Se free in order to provide a cultivation approach without Se uptake for plant growth and blank monitoring and for the validation tests (chapter 4.5.3). From each beaker 100 mL were transferred into a magentabox and 100 mL were kept for analysis. The phytoagar was cooled down for 2 hours until it formed a stable, semi-solid ground. 160 rice seeds (*Oryza sativa japonica,* cultivar *nihonmasari*) (provided by Dr. Michael Riemann, Botany Institute, KIT) were husked and sterilized by shaking for 1 min in EtOH, followed by shaking 20 min in sodium hypochlorite (NaClO) with H_2O (millipore) wash steps in between, and washed with autoclaved H_2O (millipore). 16 of them were inserted in each of the 10 magentaboxes using continuously heat-sterilized tweezers. All magentaboxes were closed and placed into a climate chamber that was adjusted to subtropical climate during rice season, 8 h of sunlight at 28°C and 16 h of darkness at 22°C with 1 h transition time each and 70 % humidity. After 16 days they were harvested and prepared according to chapter 4.3, whereby all plants of one box – respectively all roots and all shoots each – formed one sample. After proved to be Se free, the seed residuals were removed from each plant to not impact Se concentration determinations. Due to the generally low amount of sample (chapter 5.4) seeds might significantly influence the sample mass without contributing to Se amount or relevant processes. The initial phytoagar kept during preparation and the one after cultivation each was prepared according to chapter 4.2. Figure 40 shows the structural formulas of the Se species used. Figure 41 contains a scheme of the ten parallel setups (a) as well as an illustration (b) and a photograph (c) of the magentaboxes used for cultivation.

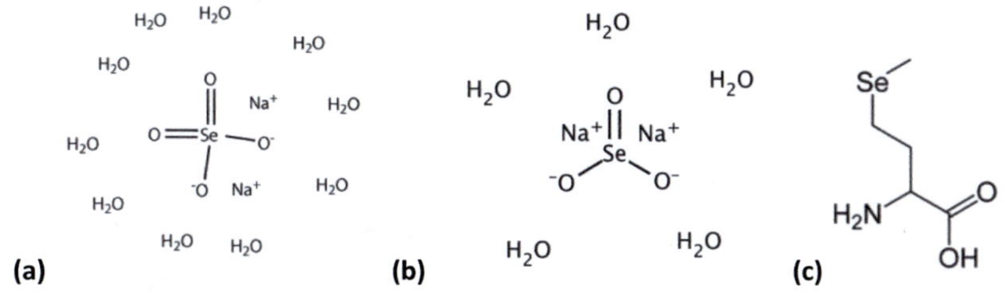

Figure 40: Molecular structure of all three Se species used – (a) sodium selenate (Na_2SeO_4 x 10 H_2O), (b) sodium selenite (Na_2SeO_3 x 5 H_2O) and (c) selenomethionine (SeMet) ($C_5H_{11}NO_2Se$).

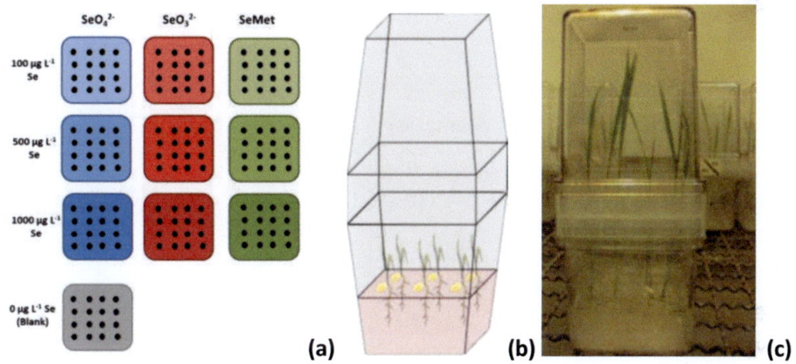

Figure 41: (a) Scheme of the experimental set up with varying Se species and concentrations (top view), (b) scheme of a magenta box with rice seedlings (Nothstein, 2015), (c) photograph of a magenta box with rice seedlings.

Minimum Parameter Experiments (MinPaX) according to this approach were repeated five times within this study, referred to as MinPax I – V. The plant growth of MinPaX I and II was monitored every second day, whereas in MinPaX III-V the plant height was determined only after harvesting. The plant samples from all experiments were weighted in total after harvesting. Heights and masses were evaluated on their influence of Se uptake and accumulation. In MinPaX I and V roots and shoots of the plants were separately prepared and analyzed, whereas in MinPaX II, III and IV the entire plant was used. All plant tissue and phytoagar samples of MinPaX I-V were prepared according to chapters 4.2 and 4.3 and analyzed on their Se concentrations (chapter 3.1.1). Phytoagar samples from MinPaX IV and V were additionally analyzed on their Se species composition (chapter 3.1.4) after cultivation to monitor Se source stability. Selected plant samples of MinPaX V were prepared in analogy to chapter 4.3 and purified according to method (C) (chapter 4.4.1). Table 16 summarizes the derived samples and analyses depending on the experiment.

Table 16: Overview over cultivation experiments performed, samples derived from them and parameters analyzed (*purified with method (C))

experiment	**plant samples**	**growth monitoring**	**Se conc.**	**Se species composition**	**δ^{82}Se**
MinPaX I	roots & shoots	✓	✓	✗	✗
MinPaX II	entire plant	✓	✓	✗	✗
MinPaX III	entire plant	✗	✓	✗	✗
MinPaX IV	entire plant	✗	✓	✓ (phytoagar)	✗
MinPaX V	roots & shoots	✗	✓	✓ (phytoagar)	✓ (plants)*

Redox stability of the source Se species in the phytoagar is mandatory for species dependent evaluation of the Se uptake into plants. Due to sterile and oxic conditions in the boxes during the entire cultivation time (Nothstein, 2015) species stability is very likely. However, for MinPaX IV and V Se species were determined in the phytoagar extracts after cultivation using the method after Bird et al. (1997) (performed by Karen Viacava and Dr. Markus Lenz, FHNW Basel) (chapter 3.1.4).
Se tends to form volatile compounds, e.g. induced by microbial or fungi activity (chapter 2.3.2). In case boxes or phytoagar were not fully sterilized, volatile losses might occur independent from plant activity. To evaluate the relevance of this factor, one "plant free" cultivation experiment was performed, other conditions being equal to MinPaX. Se concentrations were measured afterwards to quantify potential deficits.

5.2 Data processing

The raw data of Se concentration, species and isotope determination were processed according to chapters 3.1.1, 3.1.4 and 3.2.8. From Se concentrations determined, averages and standard deviations (± 1σ percentile) were calculated among MinPaX I – V for entire plants and between MinPaX I and V for roots and shoots. Averages represent general trends that can hardly be derived from single experiments due to the individual character of biological systems. Standard deviations thereby show the reliability of those trends. To quantify the major relevant processes, Se uptake, root-shoot translocation and volatilization, mass balance calculations were performed from reconstructing mass flows between compartments according to the Se distribution measured after cultivation. The amount of volatile Se being in the boxes' atmosphere (Se_{atm}) was thereby calculated as a remainder from the balance of Se initially added (Se_{ini}), Se in the cultivated plant tissue (Se_{cp}) and Se in the phytoagar after cultivation (Se_{pac}) (Equation (87)). Limitations are volatile losses not related to plants as well as errors inherent in Se determinations (chapters 3.1.1, 4.2 and 4.3).

$$Se_{atm} = Se_{ini} - Se_{cp} - Se_{pac} \tag{87}$$

Se transport between compartments

To quantify Se transport pathways within the box the system was simplified by division into three Se containing compartments phytoagar (p), cultivated plants (cp) with subcompartments roots (r) and shoots (s) as well as the atmosphere (atm). Se transport processes were defined as Se transitions between compartments. Thereby uptake (U) was simplified as the transport from phytoagar to cultivated plants, translocation (T) as the transport from roots to shoots within the plants and volatilization (V) as the transport from plant into the atmosphere. This model underlies the assumptions of unidirectional Se transport and a negligibly low amount of volatilization directly from the phytoagar V_p, which was confirmed by tests described in chapter 5.3. The sources of volatilization (roots, shoots) cannot be differentiated within this setup. Based on Terry et al. (1992) the major fraction was assumed to be volatilized via shoots. Figure 42 illustrates the scheme of compartments and transport processes.

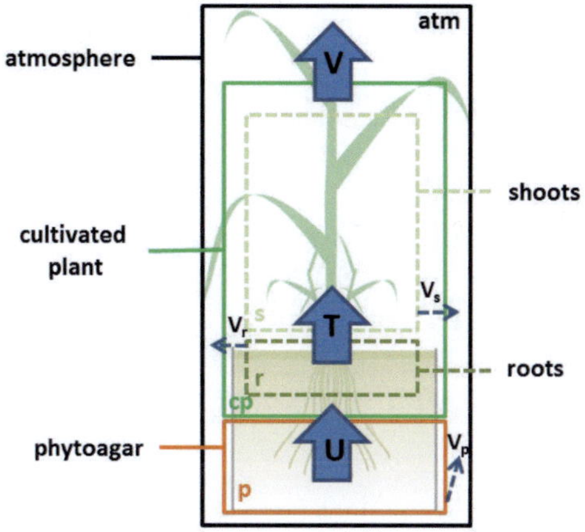

Figure 42: Mass balance model of MinPaX including compartments phytoagar (p), cultivated plants (cp) – with subcompartments roots (r) and shoots (s) – and atmosphere (atm) as well as examined transport processes uptake (U) (transport p -> cp), translocation (T) (transport r -> s) and volatilization (V) (transport cp -> atm) (large blue arrows). Volatilization from phytoagar (p -> atm) (V_p), volatilization from roots (r -> atm) (V_r) and volatilization from shoots (s -> atm) (V_s) cannot be directly derived from this setup (small broken arrows).

The Se contents [µg] initially in the phytoagar (pini) are referred to as c_{pini}, the ones after cultivation (ac) as c_{pac} and the Se contents in the cultivated plants (cp) are referred to as c_{cp} respectively c_r for roots and c_s for shoots. The sum mass flows (Q) [µg (16 days)$^{-1}$] of uptake (Q_U) and volatilization (Q_V) over the entire cultivation time were calculated according to Equations (88) and (89).

$$Q_U = c_{pini} - c_{pac} \tag{88}$$

$$Q_V = Q_U - c_{cp} = c_{pini} - c_{pac} - c_{cp} \tag{89}$$

The calculation of the translocation mass flow Q_T is more complex and less reliable than Q_U and Q_V. It should include the Se fractions volatilized from roots and shoots (Q_{Vr} and Q_{Vs}) according to Equation (90), which is hardly possible within this setup.

$$Q_T = Q_U - c_r - Q_{Vr} = c_s + Q_{Vs} \tag{90}$$

Equation (91) calculates Q_T under the assumption confirmed by Terry et al. (1992) that volatilization majorly derives from shoots. This is a reasonable approximation, but might cause errors particularly with higher Se concentrations supplied (chapter 5.5).

$$Q_T = Q_U - c_r = c_s + Q_V \tag{91}$$

To compare the extents of accumulation within compartments as well as mass flows among them relative parameters were included. Therefore accumulation (plant (a_{cp}), roots (a_r), shoots (a_s)) and uptake (u), translocation (t), volatilization (v) fractions related to the Se supplied were calculated according to Equations (92) to (99). Accumulated fractions thereby describe the extent of Se retention in system compartments, which might become important regarding nutrition and biofortification issues (chapters 2.1.2 and 2.1.4). All fractions are percentages of the Se initially supplied, which is defined as 100 %.

$$a_{pac}\ [\%] = \frac{c_{pac}}{c_{pini}} * 100 \tag{92}$$

$$a_{cp}\ [\%] = \frac{c_{cp}}{c_{pini}} * 100 = a_r + a_s \tag{93}$$

$$a_r\ [\%] = \frac{c_r}{c_{pini}} * 100 \tag{94}$$

$$a_s\ [\%] = \frac{c_s}{c_{pini}} * 100 \tag{95}$$

$$a_v\ [\%] = \frac{c_v}{c_{pini}} * 100 \tag{96}$$

$$u\,[\%] = \frac{Q_U}{c_{pini}} * 100 \tag{97}$$

$$t\,[\%] = \frac{Q_T}{c_{pini}} * 100 \tag{98}$$

$$v\,[\%] = \frac{Q_V}{c_{pini}} * 100 \tag{99}$$

Serial reactive transports might underlie rate limiting steps, describing the slowest reaction(s) that determine the overall reaction speed if dependent on each other's products. The accumulation and therefore oversupply of reagents favors the reaction of thermodynamically less stable molecules. Lighter isotopes are characterized by a slightly lower binding energy and therefore preferred in molecular transformations (Olin et al., 2005; Hoefs, 2009). Rate limiting steps tend to induce isotope fractionation, but also appear as a key process in Se transport and accumulation. Therefore revealing Se rate limiting steps in plants for the individual species is of major relevance for this study. Figure 43 illustrates the phenomenon of rate limiting steps concerning the examined processes in the cultivation system.

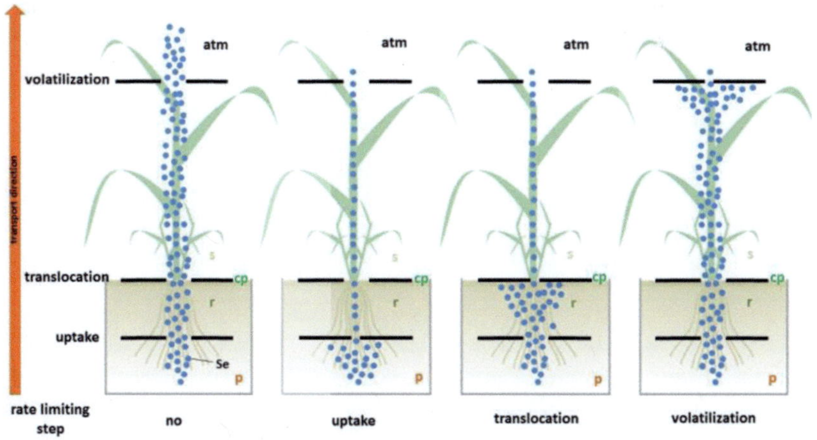

Figure 43: Illustration of potential rate limiting steps for dominant Se transport pathways (simplified) (compartments defined in analogy to Figure 42).

Se isotope fractionation in plants

For the quantification of isotope fractionation the Rayleigh model (Equation (100)) is widely applied for several stable isotope systems (e.g. Zeebe and Wolf-Gladrow, 2001) including Se (chapter 2.3.1).

$$\frac{R}{R_0} = \frac{(^{82}Se/^{76}Se)_p}{(^{82}Se/^{76}Se)_r} = f^{(\alpha-1)} \tag{100}$$

^{82}Se and ^{76}Se are isotope abundances that were measured and calculated according to chapter 3.2.8. The indices p and r represent product and reactant of the reaction regarded. f ($0 \leq f \leq 1$) is the relative fraction of Se remaining in the reactant after the considered process. α is the fractionation factor (constant) that describes the isotope fractionation characteristics of the considered process. It is the target parameter calculated from the Rayleigh equation, from which the isotope fractionation Δ is derived (Equation (9)). The knowledge of Δ for particular processes enables the reconstruction in natural environments.
The Rayleigh model was applied by Mariotti et al. 1981 and Penning (2005) on carbon and nitrogen isotope fractionation by soil microbial processes in closed systems with limited amounts of the target element (Equations (101) and (102)). Detailed derivation from Rayleigh can be found therein. The index ri represents the initial status of the reactant before the reaction regarded, p the final status of the product.

$$\delta_p = \delta_{ri} - \frac{\Delta * f * \ln(f)}{(1-f)} \Leftrightarrow \Delta = \ln(\alpha) * 1000 \Leftrightarrow \alpha = \exp(-\frac{(1-f)*(\delta_p - \delta_{ri})}{f*\ln(f)*1000} \tag{101}$$

$$\delta_r = \delta_{ri} + \Delta * \ln(f) \Leftrightarrow \Delta = \ln(\alpha) * 1000 \Leftrightarrow \alpha = \exp(\frac{(\delta_r - \delta_{ri})}{\ln(f)*1000}) \tag{102}$$

Due to similar framework conditions, the model was transferred to the cultivation system. For calculation of uptake specific α (α_U) the isotope compositions of the phytoagar before (δ_{pini}) and after cultivation (δ_{pac}) were used as initial (δ_{ri}) and final status (δ_r) (Equations (101) and (102)). δ_{pini} hereby equals the Se isotope composition measured in the Se standard solutions supplied. The fraction remaining after uptake, f_U, is defined as the Se fraction that was measured in the phytoagar after the cultivation period. α_U was then calculated according to Equation (103). Calculations of Δ were performed according to Equation (9) respectively (101) and (102).

$$\alpha_U = \exp\left(\frac{\delta_{pac} - \delta_{pini}}{\ln(f_U) * 1000}\right) \tag{103}$$

$$f_U = \frac{c_{pac}}{c_{pini}} \tag{104}$$

Using Δ_U, the hypothetical isotope ratio of the cultivated plant (product) after Se uptake δ_{cpini} under exclusion of volatilization can be derived via Equation (105) with δ_{cpini} as δ_p according to Equation (101).

$$\delta_{cpini} = \delta_{pini} - \frac{\Delta_U * f_U * \ln(f_U)}{(1 - f_U)} \tag{105}$$

δ_{cpini}, subsequently used as δ_{ri} (Equation (101)), enables the calculation of the fractionation factor for volatilization α_V based on the assumption that uptake and volatilization, which in reality take place in parallel, can be approximated as having taken place one after the other. The isotope composition measured in plant tissue after cultivation (including volatilization) was referred to as δ_{cp} and defined as the status of the reagent after reaction δ_r (Equation (102)). f_V is the fraction remaining in the plant after volatilization, it is defined as the Se amount measured in the plant related to the amount taken up, according to Equation (107). Equation (106) was used to calculate α_V.

$$\alpha_V = \exp\left(\frac{(\delta_{cp} - \delta_{cpini})}{\ln(f_V) * 1000}\right) \tag{106}$$

$$f_V = \frac{c_{cp}}{Q_U} = \frac{c_{cp}}{(c_{pini} - c_{pac})} \tag{107}$$

The fractionation by translocation α_T can only be approximated based on the assumption that volatile Se species are exclusively emitted via shoots, which is of limited validity particularly for higher concentrated Se supplementations. The fractionation induced by translocation was thereby determined via the isotope fractionation induced by volatilization and translocation combined, α_{TV}, which product the isotope composition in the shoots, δ_{shoot}, is. Using δ_{shoot} as δ_p and δ_{root} as δ_r. The initial isotope composition δ_{ri} cannot be directly measured, but calculated by rearranging Equation (102) to δ_{ri} and then insert in Equation (101) and subsequently rearrange to α (Equation (108)). The fraction remaining f_{TV} is thereby the Se amount remained in the roots related to the amount in the entire plant (Equation (109)).

$$\alpha_{TV} = \exp(-\frac{(\delta_{shoot}-\delta_{root})}{(\ln(f_T)+\frac{f_T*\ln(f_T)}{(1-f_T)})*1000}) \tag{108}$$

$$f_{TV} = \frac{c_r}{c_{cp}} \tag{109}$$

Within this setup and regarding the data given, the fractionation of translocation under exclusion of volatilization was not possible.

Against expectations, no reliable results on δ_{pac} were available for MinPaX V due to matrix specific instability and insufficient application of stabilization measures. After determining the Se concentration, phytoagar extracts were stored at 4 °C before organic destruction under unsterile conditions. This led to the formation of microbial cultures that probably changed Se composition or even formed volatile compounds, which might have impacted Se isotope composition significantly. Furthermore the precision of Se isotope determinations in phytoagar was with 1 ‰ by far lower than the one in plants and the data basis of validation was not comparably comprehensive. That is why δ_{pac} was not determined and concluded in mass balancing. For future applications, the phytoagar extracts must be treated immediately after extraction, at least concerning the destruction of organic compounds, and stored at -20 °C before purification. The precision must be improved.

However, isotope fractionation factors can be determined for uptake and volatilization in sum according to Equations (110) and (111).

$$\alpha_{UV} = -\frac{(1-f_U)*(\delta_{cp}-\delta_{pini})}{f_U*\ln(f_U)} \tag{110}$$

As all plants are exclusively products of their roots and shoots, δ_{cp} can be derived from δ_{roots} and δ_{shoots} (Equation (111)) as only isotope ratios of them were determined for MinPaX V.

$$\delta_{cp} = \frac{m_r}{m_{cp}} * \delta_{root} + \frac{m_s}{m_{cp}} * \delta_{shoot} \tag{111}$$

From the available data, the isotope fractionation induced by translocation (including volatilization) could be determined according to Equations (108) and (109).

5.3 Se source stability during cultivation

Table 17 shows the results of Se species determinations in phytoagar extracts after cultivation. The total Se concentrations before and after cultivation as well as after sample transport to Basel are given as well as Se recoveries from the postal transport (1 week, room temperature) and from species dependent and total Se determination in Basel. Finally the fraction of the Se species initially added related to the sum of Se species determined is given to represent the average species stability.

Table 17: Initial and final Se concentrations in the cultivation setups and fractions of the Se species remaining in its initial form (species measurements and data processing conducted by Lenz and Viacava, FHNW Basel (2014)) (n=2 – MinPaX IV and MinPaX V); 1)measured directly after vacuum filtration with ICP-MS 2)measured after transport (Karlsruhe – Basel) in Basel with HPLC-ICP-MS 3)total Se determined in Basel/total Se determined in Karlsruhe – effects of transport and sample preparation in Basel 4)sum Se species/total Se (Basel) – deficit is dominantly volatile species that emitted during sample preparation 5)concentration of initial Se species found in the sample/sum Se species detected (raw data available in Appendix IV, Table IV-12)

Se species added	Total source Se concentration [µg L^{-1}]1)	Total Se concentration after cultivation [µg L^{-1}]1)	Total Se concentration after cultivation [µg L^{-1}]2)	Se_{Basel}/Se_{KA} [%]3)	Σ(species)/ Se_{Basel} [%]4)	initial Se species/ Σ(species) [%]5)
selenate	119 ±0.9	3.6 ±0.2	11.2 ±2.3	303 ±46.7	n/a	
selenate	579 ±32.5	159 ±35.6	127 ±26.3	80.6 ±1.5	89.9 ±1.5	97.6 ±2.4
selenate	1190 ±20.2	978 ±24.2	727 ±48.1	74.2 ±3.1	112 ±2.6	100 ±0
selenite	114 ±3.5	76.4 ±17.4	34.9 ±1.5	48.7 ±13.0	31.9 ±13.3	96.0 ±4.0
selenite	556 ±17.0	391 ±27.1	131 ±85.1	35.2 ±24.2	54.4 ±9.0	94.3 (n=1)
selenite	1130 ±39.4	972 ±5.8	162 ±27.5	16.7 ±2.7	53.1 ±2.0	100 ±0
SeMet	108 ±1.5	26.5 ±0.4	24.2 ±2.3	91.5 ±7.4	13.0 ±0.4	96.8 (n=1)
SeMet	549 ±24.3	228 ±82.2	88.4 ±6.0	45.8 ±19.2	20.0 ±7.3	n/a
SeMet	1090 ±55.3	394 ±18.9	190 ±21.9	48.7 ±7.9	27.9 ±0.8	98.8 ±1.2

The reliability and significance of these results are limited in some respects. Comparing the Se amounts determined with ICP-MS at AGW and IC-ICP-MS respectively HPLC-ICP-MS at FHNW in Basel, high differences and Se deficits were detected regarding the total Se contents in the phytoagar extracts. The most likely reason was the accidentally long transport time from Karlsruhe to Basel due to service delay by the transport company. As mentioned in chapters 4.1 and 5.3.2.2, phytoagar extracts are unstable and serve as efficient growth medium and carbon source for microbial cultures as well, changing Se species composition or even cause volatile Se losses. Samples were unsterile after liquid vacuum extraction, and the absence of cooling might have favored microbial growth and thereby volatile losses that emitted with opening sample containers. The laboratory coworkers from Basel confirmed this assumption as they reported a typical smell of volatile organic compounds (Dr. Markus Lenz, personal comment) (Table 17 [3)]). Another limitation was the analytic facility that detected only selected Se species (chapter 3.1.4), neglecting a variety of residual Se species that might have formed during transport (chapters 2.1.4 and 2.2.4) (Table 17 [4)]). As all Se species supplied in the cultivation setups were detectable with the methods used (chapter 3.1.4), a recovery of the expected Se species related to the sum of all Se species measured was calculated and regarded as the most reasonable approach to quantify Se species stability during cultivation from the data available (Table 17 [5)]). Se yields of nearly 100 % for all species and concentrations as well as a good reproducibility indicate a certain stability of the species during cultivation. However, for reliable results this experiment should be repeated with improved framework conditions, particularly shortened transport time and cooling possibility.

Table 18 shows the results of the "plant free" cultivation experiment performed in order to monitor plant-independent volatile losses during cultivation.

Table 18: Se concentrations measured in the phytoagar extracts before* and after** "plant free" cultivation in all boxes and their Se recoveries (as the same samples were used for the methodical setup of vacuum filtration, Se_{ini} values are equal to those in Table 6 and Appendix IV, Table IV-3)

Se species added	**Se_{ini} [µg L^{-1}]* (n=3)**	**Se_{ac} [µg L^{-1}]** (n=1)**	**Se recovery [%]**	**average Se recovery species [%]**
selenate	103 ±0.7	104	101	
selenate	526 ±3.0	519	98.6	
selenate	1080 ±1.7	1080	100.5	100 ±1.0
selenite	103 ±0.5	104	101.7	
selenite	544 ±1.7	542	99.7	
selenite	1020 ±9.3	1010	98.8	100 ±1.1
SeMet	101 ±5.5	98.6	97.4	
SeMet	508 ±32.3	506	99.5	
SeMet	977 ±88.3	990	101	99.4 ±1.3
Blank	<0.1	<0.1	<0.1	
average			**99.9 ±1.2**	**99.9 ±1.2**

Se recoveries were close to 100 % with very good reproducibility among all Se species and concentrations tested. The results show that volatile losses independent from plants do not or to negligibly low extent occur during cultivation. However, one limitation of this experiment was the influence of microbial organisms located deep within the rice seeds possibly even after seed sterilization (Prof. Dr. George Alfthan, personal comment), which could not be resolved in this setup.

5.4 Growth rates and phytomass production

Figure 44 shows the average plant height development in MinPaX I and II differentiated regarding Se species and concentration. In MinPaX I the low concentrated batches of all species as well as the blank grew faster and to a higher final height. The highest concentrations of all batches grew slower with smaller final plants and even stagnating or decreasing phases. All batches had an over-proportional growth rate in the last two days before harvesting. MinPaX II did not confirm this trend. Light tendencies of lower growth rates within higher concentrated batches and vice versa were recognizable, but in general all boxes show linear growth rates that are not significantly dependent on source species or concentration. The blank box with no Se added was located in the same range concerning growth, indicating that Se addition has no significant influence on plant growth in the species and concentration ranges supplied.

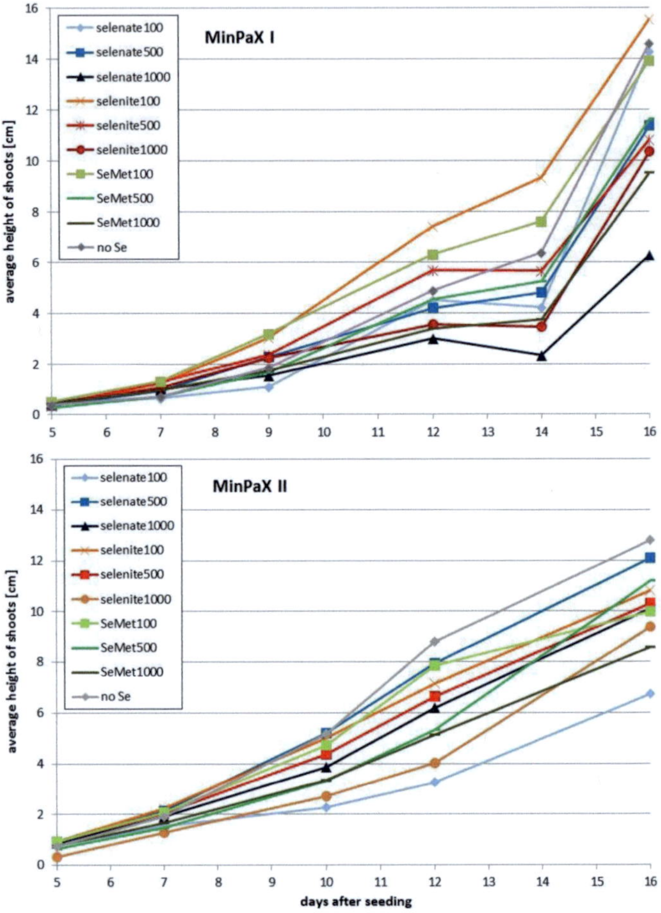

Figure 44: Growth of plants depending on Se species and concentration in MinPaX I and MinPaX II cultivation batches (uncertainty ~0.1 cm (raw data available in Appendix IV, Table IV-13).

Figure 45 shows the average length of the plants (a) and their average phytomass (b) after harvesting depending on cultivation batch for all experiments MinPaX I – V. Although individual deviations were detectable, the tendencies were similar in all cultivation experiments. The relatively high variations among MinPaX I – V and the small ones within the experiment indicate that random factors (quality of seed batch, position related to light source, plant individual development) caused differences in growth and phytomass rather than the Se source characteristic.

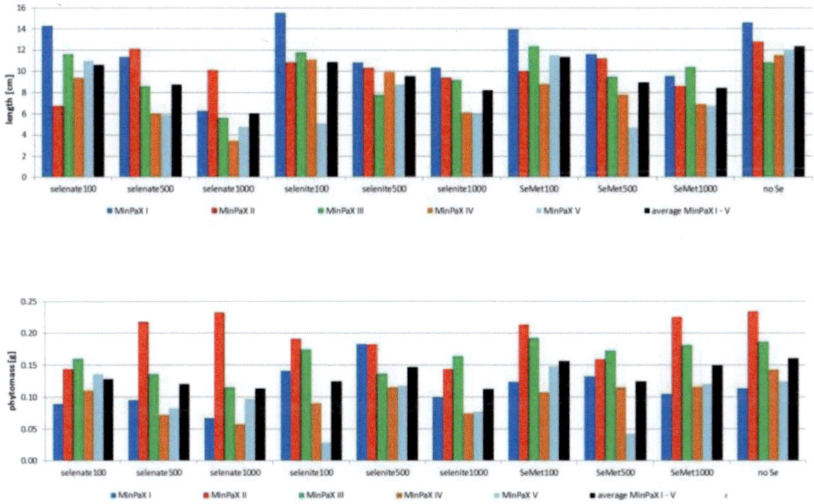

Figure 45: Average length and phytomass of plants after the cultivation period of 16 days in dependence of Se concentration and species within the different cultivation experiments (raw data available in Appendix IV, Table IV-14).

Figure 46 shows the relation between shoot length respectively phytomass and the Se fraction taken up during the entire cultivation period. No significant correlation was detectable, neither for length nor for phytomass. This confirms the indications of Figures 44 and 45 that regarding this setup Se uptake does not significantly influence plant growth and vice versa.

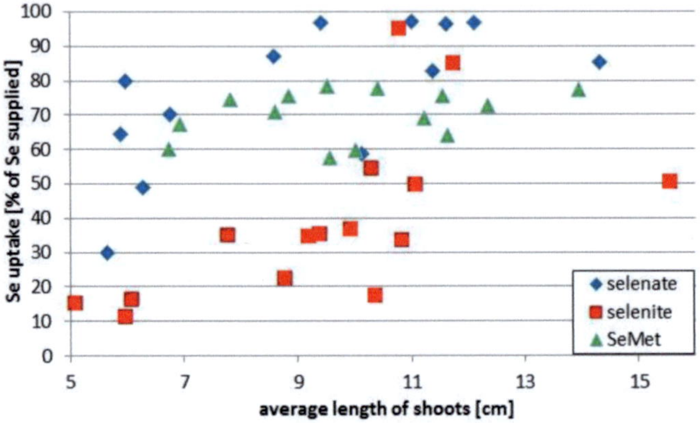

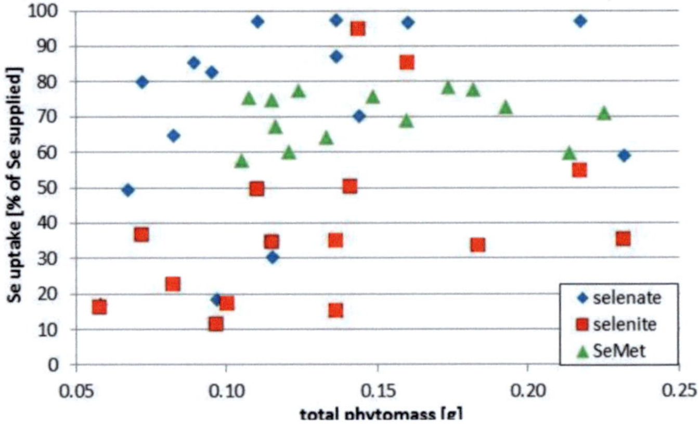

Figure 46: Relation between shoot length respectively phytomass after the cultivation period and the Se uptake depending on Se species supplied (raw data available in Appendix IV, Table IV-14 and 15).

Based on these results, the absolute Se content in the samples [µg] was preferred compared to mass related Se concentrations [µg g^{-1}] for comparison among experiments and Se source characteristics.

The absolute Se amount takes dilution effects into account, e.g. by higher individual plants and small differences in phytomass production induced by slightly higher light exposition. Furthermore, mass balancing requires the absolute amount of Se in the compartments and transport pathways. However, shoot growth, differences in plant developments (e.g. toxicity symptoms) and phytomass were monitored and related to the Se accumulation and transport processes within all experiments.

5.5 Se distribution and transport pathways

Figures 47-49 illustrate the Se distribution within the system compartments as well as the Se transport flows depending on the Se concentration supplied for the selenate setups. In the low concentrated approach of 100 µg L^{-1} initial Se concentration high uptake rates of >90 % and subsequently high translocation rates of >80 % related to the Se supplied were detected (Figure 48 and 49). Se accumulated by >50 % within the shoots were measured, which is elevated by factor 3 compared to the root-Se (Figure 47). This confirms the results of previous studies performed on selenate supplied plant cultivation presented in chapter 2.2.2. The medium concentrated approach of 500 µg L^{-1} initial Se concentration follows this trend linearly with slight decreases in uptake and subsequently in translocation (Figures 47-49). The box with the highest Se concentration shows a drop in uptake in relative (Figure 49) and absolute numbers (Figure 48). Translocation subsequently falls down (Figure 49) with the result of changing proportions of Se accumulation in roots and shoots. They are equally distributed with 10 % each (Figure 47). Under Se stress, translocation might be reduced due to toxic impacts on Se transport pathways or even actively minimized to protect the upper plant part. Se isotope signatures might reveal which mechanism caused the changing distribution.

In contrast to previous studies (chapter 2.2.4), volatilization takes place on a relatively high level for selenate supplied setups compared to other species (Figures 50-55), which equals a fraction of >20 % for the low concentrated approach. A relative decrease with rising Se supplemental concentration was detected, but different from uptake and translocation it developed linearly and with a lower slope (Figure 49). Volatilization might be a rate limiting process with a defined maximum reaction rate reached in the high concentrated setup. It therefore could cause backlogs to incorporation processes and contribute to the limitations of translocation. In this box volatilization further increased compared to the 500 µg L^{-1} setup, despite of absolute lower uptake and translocation. This indicates an active protection mechanism by the plant to avoid toxic Se levels, e.g. by shifts in the metabolic pathways in advance of higher volatilization (chapter 2.2.4, Figure 12). In this case either a higher amount of selenate was transferred into volatile organic species, which increased the number of reduction processes, or a higher amount of already reduced organic Se was converted into volatile species. The first possibility might leave enriched δ^{82}Se values especially in the shoots with a high difference to the roots as the isotopically lighter reduced Se species increasingly left the plant tissue. This is likely, because the lower Se fraction detected in the shoots within this setup (Figure 47) and the study by Terry et al. (1992) indicated a preferential volatilization pathway via shoots. Especially for selenate supply this is probable as the stability of this molecule will cause significant delays before being reduced and transformed, providing time to rush through the root system into the shoots. The second possibility could affect the roots as well, as a significantly higher fraction of Se is organically bound (Figure 11), and therefore shift the volatilization pathways in favor of the roots. In this case higher δ^{82}Se values would be detectable in both plant parts. In any case Se isotope signatures could reveal the Se related processes being responsible for changing accumulation patterns and transport flows in the high concentrated setup.

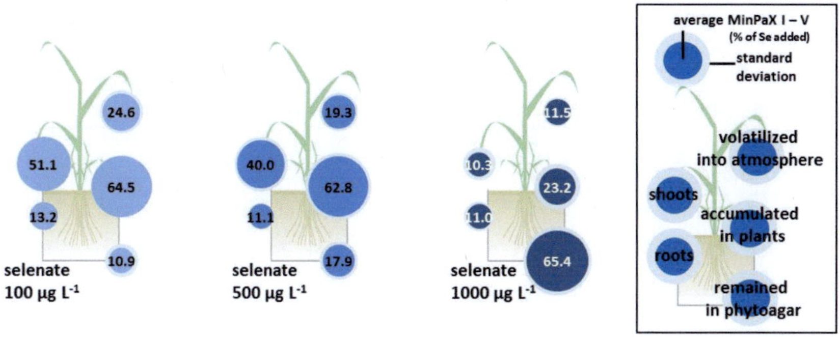

Figure 47: Distribution of Se fractions [% of Se added] within the cultivation system compartments according to Equations (92)-(96), for the selenate supplied setups (average of MinPaX I-V respectively I and V for roots and shoots) (raw data available in Appendix IV, Table IV-15).

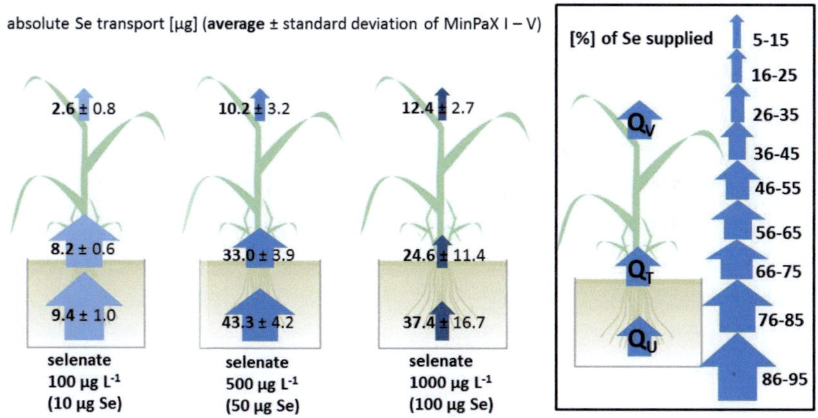

Figure 48: Absolute Se transport [µg] (according to Equations (88)-(91)) as well as relative Se transport [arrow size] (according to Equations (97)-(99)) among the cultivation system compartments for selenate supplied setups (average of MinPaX I-V respectively I and V for translocation) (raw data available in Appendix IV, Table IV-15).

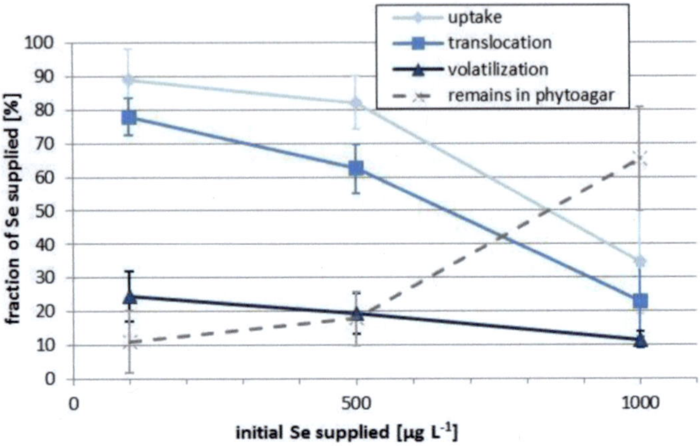

Figure 49: Relative Se transport [% of Se supplied] (according to Equations (97)-(99)) depending on the initial selenate concentration (average of MinPaX I-V respectively I and V for translocation) (raw data available in Appendix IV, Table IV-15).

Figures 50-52 illustrate the Se distribution within the system compartments as well as the Se transport flows depending on the Se concentration supplied for the selenite setups. In the lowest concentrated setup, 60 % of the Se supplied was taken up and majorly stored in the roots (Figure 50), while 10 % each were translocated and volatilized (Figure 52). This confirms previous studies on selenite uptake and translocation patters (chapter 2.2.2). Uptake decreased to a high degree and almost linearly with rising Se source concentration to 23 %. In contrast, translocation and volatilization remained at a nearly constant level, whereby volatilization was slightly elevated particularly with higher initial Se concentration (Figure 52). This linear development indicates a rate limiting step at translocation having a direct consequence on volatilization (Figure 52). An exclusive direct limitation is however unlikely as the major fraction of Se was located within the roots and furthermore organically transformed to a very high degree (chapter 2.2.2). At least a significant fraction will volatilize via roots and therefore underlies no dependence on translocation. Usually organic molecules are preferentially translocated in selenite supplied plants (chapter 2.2.2) and the formation of volatile compounds in any case requires reduction and organic transformation (Figure 12). As a consequence, the limitations of translocation and volatilization in parallel might be caused by a common prerequisite, reduction and organic transformation, instead of direct dependence. This rate limiting step might induce plant internal variations in $\delta^{82}Se$. Assuming that reduction and subsequent organic transformation exclusively occurs within the roots as indicated in chapter 2.2.2 and is limited according to Se source concentrations as assumed here, roots will be enriched in heavier Se isotopes related to Se source especially within higher concentrated setups. In contrast, shoots will be depleted in heavier isotopes related to the Se source. Amounts and pathways of volatilization will either reduce the isotopic root-shoot difference, if in analogy to selenate volatilization will majorly take place via shoots. In case it will occur majorly via the roots, $\delta^{82}Se$ variations between roots and shoots will increase. The latter is more likely because root-shoot translocation and volatilization are both on a low level and the Se amounts in the roots are much higher than in the shoots. However, Se isotope signatures might help to reveal the volatilization pathway in selenite supplied setups.

In general, Se isotope variations are expected to be much smaller than in the selenate boxes. As shown by previous studies (Figure 11), selenite is usually taken up in the inorganic form, but reduced and transformed very quickly and in major fraction minimizing detectable isotope variations. Furthermore, isotopic differences are not expected between organically bound Se and volatile organic Se species, as enzymatic processes without reduction were not reported to induce significant isotope fractionations yet (chapter 2.3.3). Differences are therefore only expected between the very small fraction of selenite and the major organic fraction. Volatilization only reduces the quantity of this major fraction and as the rates are relatively small this will probably not make a significant difference. However, Se isotope signatures might reveal the qualitative prevalence of certain volatilization pathways.

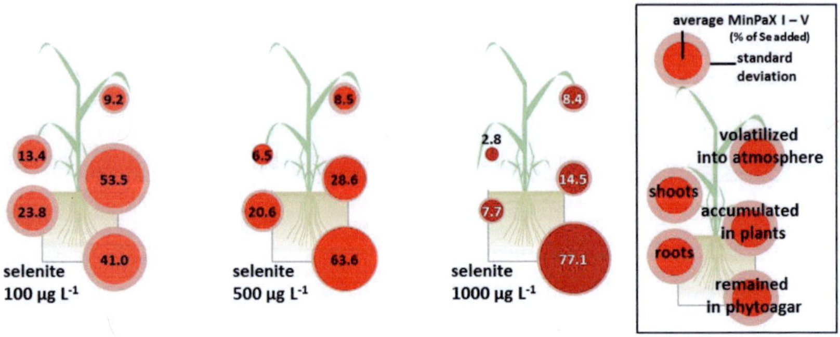

Figure 50: Distribution of Se fractions [% of Se added] within the cultivation system compartments according to Equations (92)-(96), for the selenite supplied setups (average of MinPaX I-V respectively I and V for roots and shoots) (raw data available in Appendix IV, Table IV-15).

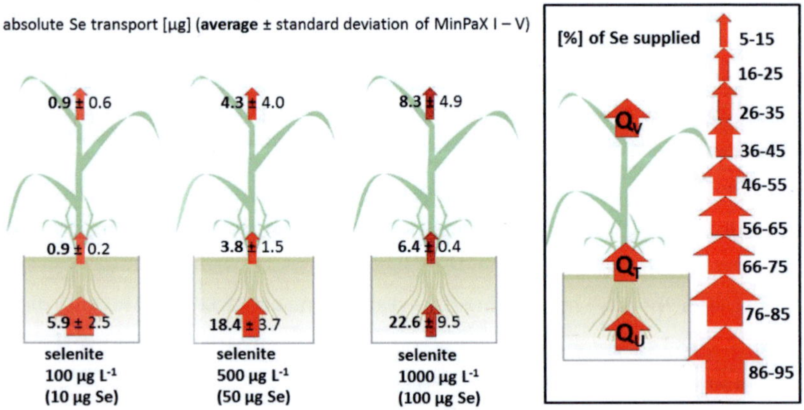

Figure 51: Absolute Se transport [µg] (according to Equations (88)-(91)) as well as relative Se transport [arrow size] (according to Equations (97)-(99)) among the cultivation system compartments for selenite supplied setups (average of MinPaX I-V respectively I and V for translocation) (raw data available in Appendix IV, Table IV-15).

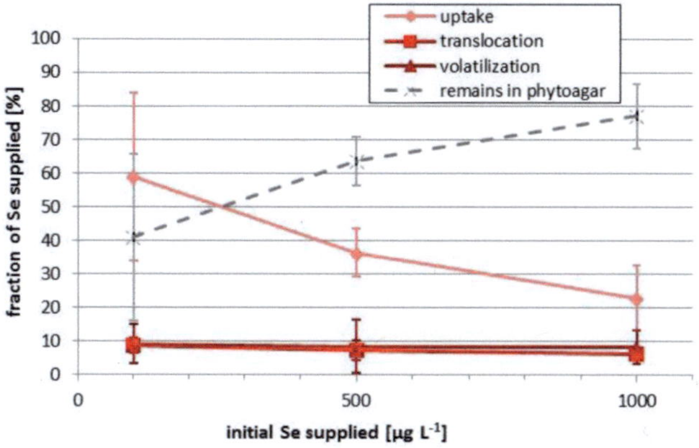

Figure 52: Relative Se transport [% of Se supplied] (according to Equations (97)-(99)) depending on the initial selenite concentration (average of MinPaX I-V respectively I and V for translocation) (raw data available in Appendix IV, Table IV-15).

Figures 53-55 illustrate the Se distribution within the system compartments as well as the Se transport flows depending on the Se concentration supplied for the SeMet setups. The lowest concentrated approach showed relatively high uptake rates of about 70 % with equal distribution among the plant parts (Figure 53). This is also confirmed by previous studies (chapter 2.2.2). While the uptake remains nearly constant with higher concentration supplied, the translocation decreased by 30 % in the 500 µg L^{-1} setups and remained constant with increasing source concentration. The transformation into volatile species might be factor limiting translocation, but with little effect. However, generally homogeneous distribution among the compartments, small process variations among concentrations supplied, low volatilization rates as well as the low tendency to induce Se isotope fractionation by enzymatic transformation only – SeMet is already available in Se(-II) and does not underly redox changes (chapter 2.2.2) – reduce the probability of insightful Se isotope variations deriving from the SeMet setups. Based on these prospects, those samples were not regarded for Se isotope analytics, as due to schedule issues only selected ones could be analyzed.

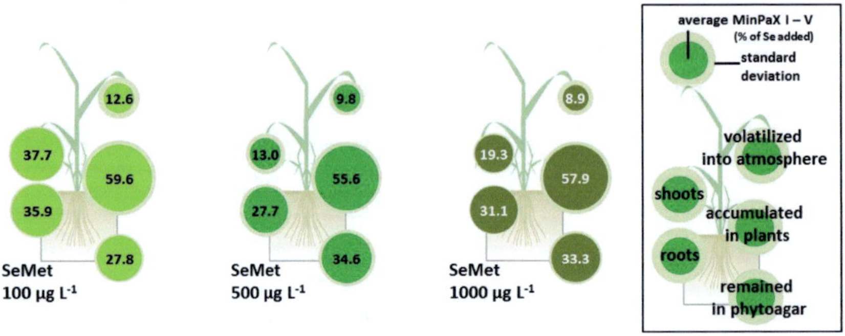

Figure 53: Distribution of Se fractions [% of Se added] within the cultivation system compartments according to Equations (92)-(96), for the SeMet supplied setups (average of MinPaX I-V respectively I and V for roots and shoots) (raw data available in Appendix IV, Table IV-15).

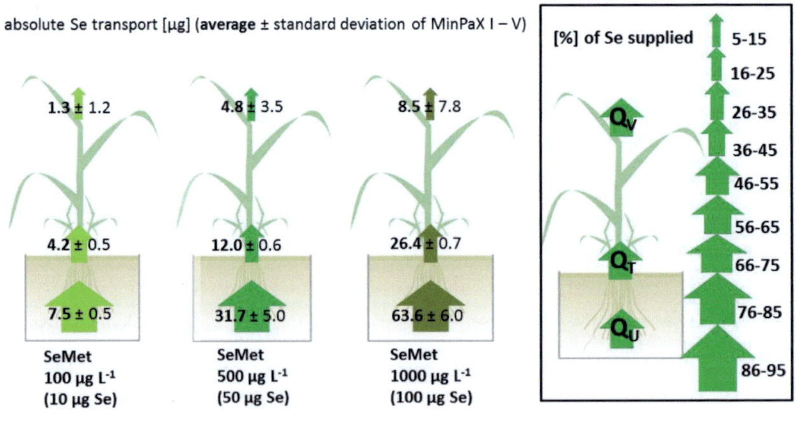

Figure 54: Absolute Se transport [µg] (according to Equations (88)-(91)) as well as relative Se transport [arrow size] (according to Equations (97)-(99)) among the cultivation system compartments for SeMet supplied setups (average of MinPaX I-V respectively I and V for translocation) (raw data available in Appendix IV, Table IV-15).

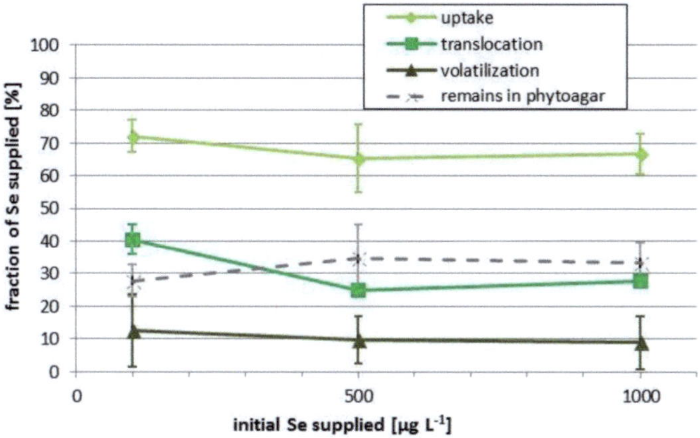

Figure 55: Relative Se transport [% of Se supplied] (according to Equations (97)-(99)) depending on the initial SeMet concentration (average of MinPaX I-V respectively I and V for translocation) (raw data available in Appendix IV, Table IV-15).

5.6 Se isotope signatures and fractionation

Plant sample treatment, purification and analytical validation enabled the precise, accurate and valid determination of Se isotope ratios ($\delta^{82}Se$) in the plant compartments roots and shoots. The Se distribution within the closed box systems and the quantification of underlying processes uptake, translocation and volatilization via mass balancing (chapter 5.5) provided the foundation to determine the process individual extent of isotope fractionation depending on Se source species and concentration. Table 19 lists the Se isotope ratios measured in roots and shoots of cultivation setups supplied with 500 µg L^{-1} and 1000 µg L^{-1} selenate as well as 500 µg L^{-1} and 1000 µg L^{-1} selenite derived from MinPaX V. These samples were chosen regarding their potential to show effects of rate limiting steps and changes in Se distribution patterns (chapter 5.5). Sodium selenate and sodium selenite standard solutions were the exclusive Se source and therefore defined as p_{ini} (chapter 5.1). These solutions were evaporated in 4M HCl to equilibrate Se species of standard and Double Spike for the determination of their Se isotope ratios. Quality control parameters defined in chapter 4.5.3 are included.

Table 19: Se isotope ratios measured and calculated in the system compartments for for cultivation batches with selenate supplied in 500 and 1000 µg L^{-1} concentrations and selenite supplied in 500 and 1000 µg L^{-1} concentrations as well as quality indicators β_{instr}, Se recoveries and residual TOC contents. $\delta^{82}Se$ values are given in relation to NIST3149 according to Equation (6). (*internal analytical error <0.1 for all samples, **fraction of Se recovered after HG and anion exchange, ***calculated from δ_{root} and δ_{shoot} according to Equation (111))

Se concentration and species supplied	System compartment	$\delta^{82}Se$ [‰]*	β_{instr}	Se recovery total [%]**	Se recovery after HG [%]	TOC [mg L^{-1}]
500 µg L^{-1} selenate	Se source (p_{ini})	-1.96	-2.02	n/a	n/a	n/a
	root	-3.19	-2.07	5.7	92.1	<0.9
	shoot	-1.95	-2.03	77.5	99.4	<0.9
	plant***	-2.36	n/a	n/a	n/a	<0.9
1000 µg L^{-1} selenate	Se source (p_{ini})	-1.96	-2.02	n/a	n/a	n/a
	root	-2.85	-2.04	47.3	87.6	<0.9
	shoot	-1.11	-2.01	91.0	91.9	<0.9
	plant***	-1.76	n/a	n/a	n/a	<0.9
500 µg L^{-1} selenite	Se source (p_{ini})	-0.49	-2.03	n/a	n/a	n/a
	root	-1.21	-2.03	64.4	95.4	<0.9
	shoot	0.28	-2.02	81.1	89.6	<0.9
	plant***	-0.37	n/a	n/a	n/a	<0.9
1000 µg L^{-1} selenite	Se source (p_{ini})	-0.49	-2.03	n/a	n/a	n/a
	root	-1.00	-2.03	49.9	97.7	<0.9
	shoot	-0.07	-2.01	66.8	96.1	<0.9
	plant***	-0.51	n/a	n/a	n/a	<0.9

The $\delta^{82}Se$ values determined in the plant samples covered a range of -3.2 ‰ to 0 ‰, which is characteristic for natural systems (Figure 14). β_{instr} shows stable values with low scattering and no detectable dependencies on $\delta^{82}Se$. Although total Se recoveries including HG and anion exchange underlay relatively high variations, Se recoveries of HG only were constantly on a high level above 88 %. This indicates a high precision of at least 0.4 ‰ for $\delta^{82}Se$ according to Figure 38. No TOC could be detected in any sample determined. All of these parameters indicate high precision and validity for $\delta^{82}Se$ values, making the data basis high extent reliable (chapter 4.5.3).
Table 20 shows the Se isotope differences between compartments and the isotope fractionation characteristics of Se uptake and translocation occurred within the same cultivation setups. The fractions remaining f_U and f_{TV} as well as the Se contents used for calculation were taken from the individual MinPaX V data.

Table 20: Se isotope differences and fractionation characteristics for uptake and translocation each including volatilization effects for cultivation batches with selenate supplied in 500 and 1000 µg L^{-1} concentrations and selenite supplied in 500 and 1000 µg L^{-1} concentrations (δ represents the respective $\delta^{82}Se$) (raw data available in Table 19 and Appendix IV, Table IV-15)

Process	**Parameter**	**Se concentration and species supplied**			
		500 µg L^{-1} selenate	**1000 µg L^{-1} selenate**	**500 µg L^{-1} selenite**	**1000 µg L^{-1} selenite**
Uptake (incl. volatilization)	δ_{cp}- δ_{pini} [‰]	-0.40	0.20	0.12	-0.02
	f_U	0.39	0.95	0.84	0.97
	α_{UV}	0.9993	1.0002	1.0001	0.9999
	Δ_{UV} [‰]	-0.67	0.21	0.13	-0.02
Translocation (incl. volatilization)	δ_{shoot}- δ_{root} [‰]	1.24	1.74	1.49	0.93
	c_r	0.14	0.25	0.34	0.17
	c_{cp}	0.87	0.45	0.55	0.28
	f_{TV}	0.16	0.56	0.61	0.62
	α_{TV}	1.0035	1.0024	1.0019	1.0012
	Δ_{TV} [‰]	3.51	2.34	1.92	1.19

Isotope differences between initial and plant Se were relatively low in every system regarded. The 500 µg L^{-1} selenate box showed a low, but significant depletion in ^{82}Se of plant tissue related to the Se source. The calculated fractionation -0.67 ‰ is less than the fractionation of -1.1 ‰ between Se source and plant reported by Herbel et al. (2002), but follows the same trend of slight depletion. Se uptake occurs via specific transport channels for selenate and selenite each and without changing molecular composition (chapter 2.2.2). Furthermore the molecules are too large to dissolve kinetic isotope fractionation caused by diffusion or dispersion (Hoefs, 2009). Therefore uptake itself could unlikely cause isotope fractionation. As Herbel et al. (2002) took in situ plant samples from a wetland, the Se source available probably consisted of a Se species mixture and the plant might have taken up certain species preferentially, e.g. isotopically lighter Se_{org} when available as SeMet (chapter 2.2.2). In this study the Se sources were homogeneous regarding species. Rhizospheric processes changing pH and redox conditions could have taken place, e.g. organic acid chelation (e.g. Rauland-Rasmussen et al., 2008), and might have reduced minor fractions of selenate prior to uptake. However, as this fractionation was very small and the one in the other boxes not detectable despite of high analytic precision, Se isotope fractionation at uptake can be regarded as not significant. This indicates that, not as expected, volatilization does not play a major role regarding Se isotope variations. Emission of volatile species from the plants does not include redox changes and only reduces the lighter Se_{org} fraction (chapter 2.2.4), which extent is not high enough to cause changes in isotope composition. Therefore volatilization pathways are hardly differentiable using Se isotope signatures in this setup. Gas phase trapping from the box and direct analyzation of the volatile species could on the other hand provide a deeper insight into volatilization characteristics. Se isotope variations between root and shoot were detected to be by far higher than the analytical precision reached for plant samples, which applies for all setups examined. Selenate setups underlay higher fractionation than selenite ones, which was caused by differing Se species composition between plant parts. Several studies previously showed that the organic Se fraction is much higher in the roots than in the shoots. This particularly applies for selenate supplementation, as a small fraction is reduced to Se_{org} within the roots, whereas the fraction remaining selenate is translocated to the shoots (chapter 2.2.2). The reduction step prefers light isotopes that were thereby fixed in the root tissue, when remaining heavier isotopes move on to the shoots causing fractionation of +3.5 ‰ for the supplementation of 500 µg L^{-1} Se (chapter 2.3). The lower fractionation at the 1000 µg L^{-1} setup was likely due to the lower translocation rate. An active mechanism to avoid Se translocation as suspected could not be observed in the data available. This mechanism might exist anyway, but not cause Se isotope fractionation. On the other hand, translocation rates might have been impacted directly by Se poisoning effects. Such effects were detected by Nothstein (2015) at supplied concentrations above 1000 µg L^{-1} applying the same setup. The assumption that volatilization was a rate limiting step for Se translocation in selenate supplied plants could not be confirmed. Figure 56 illustrates the processes that could have occurred and the Se isotope fractionation presumably induced by them.

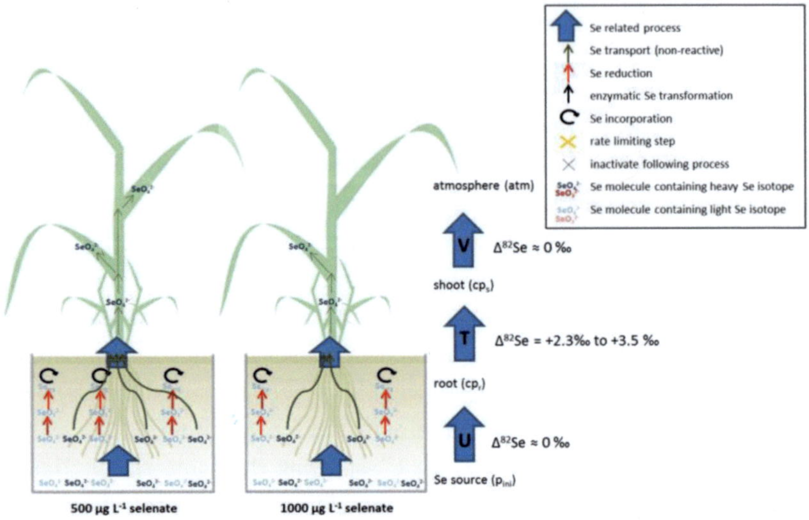

Figure 56: Se isotope fractionation and presumed underlying processes in the case of selenate supply (raw data available in Table 19 and Appendix IV, Table IV-15).

The selenite setups showed a similar trend regarding translocation, but significantly less pronounced than detected in selenate supplied plants. However, translocation led to Se isotope fractionation of almost +2 ‰. In analogy to selenate, a certain fraction of selenite probably remained in this species and was preferentially translocated, whereas the reduced organic fraction mainly remained in the roots. These results differ from a study by Kahakachchi et al. (2004), who reported that selenite was quickly and by >90 % reduced and organically bound in the roots. Only a minor fraction was translocated, available as Se_{org}. Kahakachchi et al. (2004) studied *Brassica juncea*, which is characterized as secondary Se accumulator in contrast to the non-accumulator *Oryza sativa* examined in this study. Mounicou et al. (2006) showed that secondary accumulators contain up to 50 % water soluble organic Se forms that can easily be translocated (e.g. SeMet, also chapter 5.5). In contrast, non-accumulators include high proportions of Se bearing proteins, large molecules that consist of various soluble organic Se species and tend to be incorporated into the plant tissue rather than translocated (chapter 2.2.3). Thereby Se_{org} rather remains in the roots within *Oryza sativa* and a significant fraction of selenite is translocated instead inducing enrichment in light isotopes in the roots. Nothstein (2015), who quantified Se species within *Oryza sativa* supplied with selenite, found on average 28 % Se that was still available as selenite in the shoots, compared to 7 % being selenite in the roots, the rest being Se_{org}. These findings confirm the assumption that significant amounts of Se were translocated as selenite as presumed from Se isotope fractionation. Thus, particular metabolic pathways and the character of organic Se species can be detected in plants using isotope signatures, although these cannot directly differentiate between organic Se species. In the 1000 µg L^{-1} supplied box, the fractionation was with +1.2 ‰ significantly lower than in the 500 µg L^{-1} one, although translocation rates were nearly equal. Evaluating Se transport processes (chapter 5.5) a rate limiting step was revealed concerning translocation and volatilization. Reduction rates itself might have reached a limitation causing an accumulation of heavier selenite within the roots. Mechanisms involved in Se translocation itself might also have reached limits reducing the transfer of selenite and thereby enrichment of heavy isotopes in the shoots. Furthermore, the formation rate of large organic Se compounds (e.g. Se bearing proteins) from soluble organic Se molecules (e.g. SeMet) could have been limited as well. This might have led to accumulation of soluble organic Se in the roots and, as these compounds tend to translocate (chapter 5.5), to a transfer into the shoots. Thereby the total organic Se fraction would have increased in the shoots and decreased in the roots, leading to a smaller difference in $\delta^{82}Se$ compared to a lower Se concentration supplied. All of the reactions described might play a significant role in the occurrence of Se isotope variations. Figure 57 illustrates processes presumed to have taken place and the Se isotope fractionation detected.

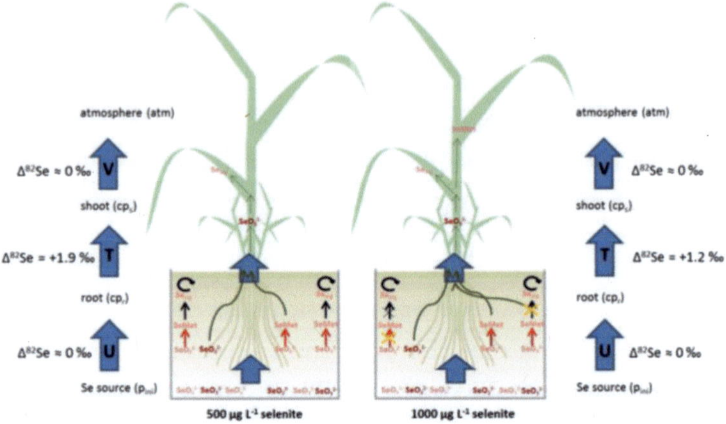

Figure 57: Se isotope fractionation and presumed underlying processes in the case of selenite supply (SeMet exemplarily represents soluble organic Se species, e.g. amino acids; Se_{org} represents large organic molecules tending to incorporation, e.g. proteins) (symbol keys in Figure 56) (raw data available in Table 19 and Appendix IV, Table IV-15).

5.7 Potential applications

Distribution of Se within plant compartments by trend confirmed previous studies presented in chapter 2.2.2 at low concentration supplied. However, this particular setup revealed the sensitivity of Se distribution and transport pathways on Se concentrations. Selenate, the species previously identified as the one with the highest potential for plant biofortification (Longchamp et al., 2015) and phytoremediation (Banuelos and Lin, 2005), underlies certain limitations concerning concentrations. Thereby volatilization via the shoots plays a key role that might become relevant for those applications. In contrast, selenite transport is already limited by uptake and inhibits translocation that cannot exceed critical limits in the upper, edible plant parts. This phenomenon might be used to monitor and avoid Se excess in plant foods if cultivated on soils or with water highly enriched in Se. SeMet was shown to be taken up at similarly high rates in a wide concentration range and tends to a homogeneous distribution within the entire plant tissue. This might be applied in long-term biofortification projects as a certain part is transported to the edible upper plant parts and the residual fraction is bound to the roots and thereby retained in the soil for subsequent plant cultivation. This retained fraction might serve as a continuous Se source within the soil. In contrast, supplemented selenate is either majorly removed from the fields with crop harvest or underlies leaching processes and can therefore unlikely be a predictable and long-term stable Se source (Alfthan et al., 2015). However, particularly higher concentrations revealed specific rate limiting steps that significantly influence Se transport and accumulation depending on the Se species supplied. Plant internal Se distributions and the processes inducing their extent and changes can be traced with Se isotope signatures. Root-shoot translocation leads to high isotope fractionation due to varying fractions of Se species stored in the roots respectively translocated to the shoots. Thereby translocation is the key process for monitoring Se accumulation in edible plant parts and its species distribution therein. This in turn defines Se bioavailability and furthermore the potential of uptake, storage and volatile emission, which is crucial for phytoremediation applications (Banuelos and Lin, 2005). Transferring this knowledge to in situ plants, the determination of Se isotope signatures might enable the detection of rate limiting steps without complex and lengthy cultivation experiments or long-term in situ monitoring. Thereby the causes of Se related problems, e.g. lacking accumulation in grains and fruits despite of high Se supplementation or Se excess in plants despite of moderate soil concentrations, could be indicated by determining a single state parameter. This makes Se isotope analyses a powerful tool to find solutions on Se related challenges, particularly in agricultural systems. The interpretation of Se isotope signatures determined in in situ sampled plants must of course include influence factors beyond plant related processes (e.g. soil properties, soil solution chemistry, microbial activity). However, these factors control Se properties outside the plants. After entering the plant, Se underlies processes that do not significantly differ from those reflected in the *Minimum Parameter* approach.

6 Conclusions and outlook

6.1 Evaluation of goals

A major goal of this study was the evaluation of Se distribution as well as Se isotope variations in the soil-plant system. To separately examine plant related processes in that regard, a *Minimum Parameter* approach was developed. This excluded the influence of soil characteristics, nutrients and solutes as well as microbial activity. Basic prerequisites for the evaluation of Se distribution are reliable and standardized sample preparation procedures for plant and phytoagar tissue. Based on these, purification procedures were applied that separate sample-Se from matrix compounds. After development, examination and assessment of diverse methods an innovative microwave digestion procedure was chosen for plant tissue preparation and a newly developed vacuum filtration method for phytoagar material. Both guaranteed full Se recovery. Organic destruction was fully enabled by microwave digestion, but limited for phytoagar filtration although being on a high level. This might be improved e.g. by longer reaction times or a different acid composition. Higher temperatures would very likely lead to uncorrectable Se losses. Future studies can build on these methods and use them as standard procedures for concentration measurements, isotope determinations or examination of other sample characteristics (e.g. molecular composition of phytoagar). The methods tested and optimized for sample purification all underlay characteristic strength and limitations. The efficiency of anion exchange depended on the matrix particularities to high extend. Thiol retention was robust and reliable with high Se recoveries for all matrices used. Both methods were able to reduce matrix elements to a minimum, but retained significant amounts organic compounds in the purified phases. Hydride separation enabled a full purification of samples from their organic matrices, but revealed limitations concerning high metal contents. Validation tests confirmed the high impacts of organic residuals, yielding invalid results for samples treated with anion exchange and thiol retention. Hydride separation gained valid and reproducible results with a precision of 0.2 (±0.2) ‰ for plants and 1.1 (±0.1) ‰ for phytoagar. This study revealed the actual sensitivity of Se isotope analytics towards organic compounds. It therefore provides an exclusive method that is suitable for Se isotope analyses in plants and other organic rich samples. If available in samples, high metal impacts could be mitigated by previous anion exchange. Phytoagar precision might be improved by higher organic destruction rates prior to purification. For samples with no significant organic compounds, anion exchange might be the preferential purification method. In contrast, despite its reliably high Se recoveries among a variety of sample types thiol retention in general is hardly suitable for valid Se isotope determinations. High amounts of organic residuals that are mobilized from the cellulose powder will very likely lead to organic matrix effects even if the samples themselves do not contain significant organic fractions. However, this method might be used for pre-concentration of Se poor samples due to its high Se binding capacity as tested with a similar setup by Savard et al. (2006).

The *Minimum Parameter* cultivation setup was proved to exclude detectable microbial activity and to keep Se sources stable during the entire cultivation period. Plant growth and development was neither significantly promoted nor retarded by the supplementation of Se, cultivation setups were highly reproducible regarding the individuality of biological organisms. The procedure developed for organic sample treatment yielded sufficient analytical precision to detect plant internal Se isotope variations. From the methodical perspective this approach proved to be suitable for differentiated examination of Se related processes in plants. Se uptake and translocation in low and medium concentrated setups occurred in analogy to previously published studies. New insights into plant Se metabolism were gained concerning volatilization at all source concentrations as well as uptake, translocation and volatilization at the highest supplied Se concentrations among all species. Despite of high rates, uptake and volatilization did not induce detectable Se isotope fractionation. In contrast, high fractionations related

to translocation revealed plant internal Se species distribution, which turned out to shift with rising Se source concentrations supplied.
In summary, this thesis provides a comprehensive, reliable method for organic sample treatment and valid Se isotope analyses in plants. Furthermore a cultivation setup was developed and evaluated meeting the demands of differentiated process investigations and providing the opportunity of extensions to successive approximation to natural conditions. First new insights and approaches for the investigation of the Se cycle in plants using Se isotope variations were performed. Characteristic relations between single metabolic processes and isotope signatures were discovered. However, to strengthen the reliability of those results, the experiments have to be repeated additionally including Se isotope determinations in phytoagar. This broader data basis is necessary due to the biological individuality of plants.

6.2 Next steps

The present thesis shows that the *Minimum Parameter* approach enables the differentiated examination of plant related processes via mass balancing and Se isotope determinations. Based on these, this methodical setup offers several possibilities for extensions.
The trapping and analysis of the gas phase for Se concentration and Se isotope signatures could be a promising approach to investigate the volatilization process more thoroughly and to increase precision and reliability of the mass balance model. This was already applied for Se isotope applications in microbial, fungi and algae cultures (Johnson et al., 1999; Schilling et al., 2011b; Schilling et al., 2013) and might be transferable to plant cultures if volatilization rates are sufficiently high and the system remains sterile.
The influence of soil composition and soil solution chemistry can be simulated by extending the *Minimum Parameter* approach on those components. Nothstein (2015) successfully developed and implemented cultivation experiments in equal boxes using optimum nutrient solution that included ubiquitary anions and cations as well as artificial soil consisting of quartz,

kaolinite and goethite. These setups could be used for Se isotope determinations as well, although modifications in sample treatment and purification might be necessary. Further extensions might be the addition of Se fertilizer to simulate biofortification measures or the use of altered nutrient solution to simulate pH and redox changes that result from particular agricultural activity, which proved to be a critical factor in the Se cycle. Plant species might be varied as well, e.g. dependent on the type of Se accumulator or related to their relevance for global food security.

The *Minimum Parameter* setup and its extensions are very suitable to investigate the role of plants and their interaction with some factors, but it is limited in reflecting entire ecosystems. Inclusion of organic phases is hardly feasible, because microbial activity will lead to oxygen insufficiency and CO_2 excess in the closed system, thereby changing framework conditions. In addition to that it is impossible to grow plants to maturity for an investigation of their grains and fruits, which would be highly relevant with regard to nutrition issues. Closed box systems also do not allow interventions during cultivation time, e.g. fertilizer addition or sampling. For this purpose greenhouse studies could serve as a transition to natural conditions: they enable (to a limited extend) a controlled growth and the inclusion of organic phases at the same time. Results derived from these setups could be verified in natural systems that are well known regarding their Se cycle like Se accumulated agricultural fields in Punjab (India) or wetlands suffering from Se excess in Kesterson Reservoir, California.

7 References

AHF (2014): Characteristics of PFA beakers and equipment. AHF, Tübingen (https://www.ahf.de/pdf/PFA-Laborprodukte/PFA-Labware.pdf) (last download 05.12.2015).

Alfthan G., Eurola M., Ekholm P., Venäläinen E.-J., Root T., Korkalainen K., Hartikainen H., Salminen P. Hietaniemi V., Aspila P., Aro A. (2015): Effects of nationwide addition of selenium to fertilizers on foods, on animal and human health in Finland: From deficiency to optimal selenium status of the population. Journal of Trace Elements in Medicine and Biology 31, 142-147.

Antonelli M., Siciliano G., Turvani M.E., Rulli M.C. (2015): Global investments in agricultural land and the role of the EU: Drivers, scope and potential impacts. Land Use Policy 47, 98-111.

Arnold T., Markovic T., Kirk G.J.D., Schönbächler M., Rehkämper M., Zhao F.J., Weiss D.J. (2015): Iron and zinc isotope fractionation during uptake and translocation in rice (*Oryza sativa*) grown in oxic and anoxic soils. Comptes Rendus Geoscience (in press).

Bajaj M., Eiche E., Neumann T., Winter J., Gallert C. (2011): Hazardous concentrations of selenium in soil and groundwater in North-West India. Journal of Hazardous Materials 189, 640-646.

Banuelos G.S. and Lin Z.Q. (2005): Phytoremediation management of selenium-laden drainage sediments in the San Luis Drain: a greenhouse feasibility study. Ecotoxicology and Environmental Safety 62, 309-316.

Baum M.K., Campa A., Miguez-Burbano M.J., Burbano X., Shor-Posner F. (2001): Role of selenium in HIV/AIDS. In: D.L. Hatfield (Ed.): Selenium: Its Molecular Biology and Role in Human Health, 247-256. Boston Kluer Academic Publishers.

Bell P.F., Parker D.R., Page A.L. (1992): Contrasting Selenate-Sulfate Interactions in Selenium-Accumulating and Nonaccumulating Plant Species. Soil Science Society of America Journal 56.

Berglund M. and Wieser M.E. (2011): Isotopic compositions of the elements 2009 (IUPAC Technical Report). Pure & Applied Chemistry 83 (2), 397-410.

Bio-Rad (2011): AG 1, AG MP-1 and AG 2 Strong Anion Exchange Resin – Instruction Manual. (available on request).

Bird S.M., Honghong G., Uden P.C., Tyson J.F., Block E., Denoyer E. (1997): High performance liquid chromatography of selenoamino acids and organo selenium compounds. Speciation by inductively coupled plasma mass spectrometry. Journal of Chromatography A 789, 349-359.

Bodik I., Lyman W., Reehl W.F., Rosenblatt D.H. (1988): Environmental Inorganic Chemistry – Properties, Processes and Estimation Methods. Pergamon Press 1988.

Brinkman M., Reulen R. C., Kellen E., Buntinx F., Zeegers, M. P. (2006): Are men with low selenium levels at increased risk of prostate cancer? European Journal of Cancer 42, 2463–2471.

Campbell A.D. (1992): A critical survey of hydride generation techniques in atomic spectroscopy (IUPAC Technical Report). Pure & Applied Chemistry 64 (2), 227-244.

Cappa C.D., Hendricks M.B., DePaolo D.J., Cohen R.C. (2003): Isotopic fractionation of water during evaporation. Journal of Geophysical Research 108, NO.D16.

Carignan J. and Wen H. (2007): Scaling NIST SRM 3149 for Se isotope analysis and isotopic variations of natural samples. Chemical Geology 242, 347-350.

CGIAR – Global Rice Science Partnership (2013): The Rice Almanac. Source Book for One of the Most Important Economic Activities on Earth. Fourth Edition. (http://books.irri.org/9789712203008_content.pdf) (last download 05.12.2015).

Chakraborty S., Bardelli F., Charlet, L. (2010): Reactivity of Fe(II) on calcite: Selenium reduction. Environmental Science & Technology 44, 1288–1294.

Chemotrade Chemiehandelsgesellschaft mbH (2012): Certificate of Analysis for 77-Se and 74-Se – costumer order (available on request).

Chilimba A.D.C., Young S.D., Black C.R., Rogerson K.B., Louise Ander E., Watts M.J. (2011): Maize grain and soil surveys reveal suboptimal dietary selenium intake is widespread in Malawi. Scientific Reports 1:72.

Clark L.C., Combs G.F., Turnbull B.W., Slate E.H., Chalker D.K., Chow J. (1996): Effects of selenium supplementation for cancer prevention in patients with carcinoma of the skin. A randomized controlled trial. Journal of American Medical Association 276, 1957-1963.

Clark S.K. and Johnson T.M. (2008): Effective Isotope Fractionation Factors for Solute Removal by Reactive Sediments: A Laboratory Microcosm and Slurry Study. Environmental Science & Technology 42, 7850-7855.

Compston W., Oversby V.M. (1969): Lead Isotopic Analysis Using a Double Spike. Journal of Geophysical Research 74 (17), 4338-4348.

Contempre B., Duale N. L., Dumont J. E., Ngo B., Diplock A. T., Vanderpas J. (1992): Effect of selenium supplementation on thyroid hormone metabolism in an iodine and selenium deficient population. Clinical Endocrinology (Oxford, U.K.) 36 (6), 579–583.

Cussler E.L. (1997): Diffusion - Mass Transfer in Fluid Systems. Cambridge University Press, Cambridge/New York, ISBN 0521564778.

Davies E., Huang Y., Harper J.B., Hook J.M., Thomas D.S., Burgar I.M., Lillford P.J. (2010): Dynamics of water in agar gels studied using low and high resolution 1H NMR spectroscopy. International Journal of Food Science and Technology 45, 2502-2507.

De Souza M.P., Pilon-Smits E.A.H., Lytle C.M., Hwang S., Tai J., Honma T.S.U., Yeh L. and Terry N. (1998): Rate-limiting steps in selenium assimilation and volatilization by Indian mustard. Plant Physiology 117, 1487–1494.

Deratani A., Sebille B., Hommel H., Legrand A.P. (1983): Reactivity and structure of the complex between thiol groups of a chelating resin and a copper ion. Reactive Polymers 1, 261-266.

Dhillon K. S., Dhillon S. K., Pareek N. (2005): Distribution and bioavailability of selenium fractions in some seleniferous soils of Punjab, India. Archives of Agronomy and Soil Science 51 (6), 633 — 643.

Dhillon K. S. and Dhillon S. K. (2003): Distribution and management of seleniferous soils. Advanced Agronomy 79, 119–184.

Dodson M.H. (1963): A theoretical study of the use of internal standards for precise isotopic analysis by the surface ionization technique: part I – general first-order algebraic solutions. Journal of Scientific Instruments 40, 289-295.

Druschel G.K., Schoonen M.A.A., Nordstrom D.K., Ball J.W., Xu Y., Cohn C.A. (2003): Sulfur geochemistry of hydrothermal waters in Yellowstone National Park, Wyoming, USA. III. An anion exchange resin technique for sampling and preservation of sulfoxyanions in natural waters. Geochemical Transactions 4(3), 12-19.

Duchefa Biochemie B.V. (2011): Phytoagar safety data sheet. (www.duchefa-biochemie.com/msds/download/id/1450) (last download 05.12.2015).

Eichhorn V. (2015): Optimierung und Anwendung von Selen-Separationsverfahren mittels Säulentrennung für Selenisotopenmessungen an Bodenproben. Approved M.Sc. thesis, Institute of Applied Geosciences, Karlsruhe Institute of Technology.

El Mehdawi A.F., Quinn C.F., Pilon-Smits E.A.H. (2011): Selenium Hyperaccumulators Facilitate Selenium-Tolerant Neighbors via Phytoenrichment and Reduced Herbivory. Current Biology 21, 1440-1449.

Ellis D.R. and Salt D.E. (2003): Plants, selenium and human health. Current Opinions in Plant Biology 6, 273–279.

Ellis A.S., Johnson T.M., Herbel M.J., Bullen T.D. (2003): Stable isotope fractionation of selenium by natural microbial consortia. Chemical Geology 195, 119-129.

Elwaer N. and Hintelmann H. (2007): Comparative performance study of different sample introduction techniques for rapid and precise selenium isotope ratio determination using multi-collector inductively coupled plasma mass spectrometry (MC-ICP/MS). Analytical and Bioanalytical Chemistry 389, 1889-1899.

Elwaer N. and Hintelmann H. (2008a): Comparing the precision of selenium isotope ratio measurements using collision cell and sector field inductively coupled plasma spectrometry. Talanta 75, 205-214.

Elwaer N. and Hintelmann H. (2008b): Precise selenium isotope ratio measurement using a multimode sample introduction system (MSIS) coupled with multicollector inductively coupled mass spectrometry (MC-ICP-MS). Journal of Analytical Atomic Spectrometry 23, 1392-1396.

Elwaer N. and Hintelmann H. (2008c): Selective separation of selenium (IV) by thiol cellulose powder rand subsequent selenium isotope determination using multicollector inductively coupled plasma mass spectrometry. Journal of Analytic Atomic Spectrometry 23, 733-743.

Fan A.M., Book S.A., Neutra R.R., Epstein D.M. (1988): Selenium and human health implications in California's San Joaquin Valley. Journal of Toxicology and Environmental Health 23 (4), 539-59.

Fernández-Martínez A. and Charlet L. (2009): Selenium environmental cycling and bioavailability: A structural chemist point of view. Reviews in Environmental Science and Bio/Technology 8, 81–110.

Fitzpatrick A.J., Kyser T.K., Chipley D. (2009): Selenium measurement in sulphide by hydride generation high-resolution inductively coupled plasma mass spectrometry. Geochemistry: Exploration, Environment Analysis 9, 93-100.

Fordyce F.F. (2007): Selenium geochemistry and health. Ambio 36, 94–7

Franck R.R. (2005): Bast and other plant fibres. Woodhead, Cambridge UK, ISBN 1855736845.

Gailer J. (2007): Arsenic–selenium and mercury–selenium bonds in biology. Coordination Chemistry Reviews 251, 234–254.

Gao X., Root R.A., Farrell J., Ela W., Chorover J. (2013): Effect of silicic acid on arsenate and arsenite retention mechanisms on 6-L ferrihydrite: A spectroscopic and batch adsorption approach. Applied Geochemistry 38, 110–120.

Gissel-Nielsen G, Gupta UC, Lamand M, Westermarck T. (1984): Selenium in soils and plants and its importance in livestock and human nutrition. Advances in Agronomy 37, 397–460.

Glaubitz U., Erban A., Kopka J., Hincha D.K., Zuther E. (2015): Metabolite Profiling Reveals Sensitivity-Dependent Metabolic Shifts in Rice (Oryza sativa L.) Cultivars under High Night Temperature Stress. Agriculture and Climate Change – Adapting Crops to Increased Uncertainty (AGRI 2015). Procedia Environmental Sciences 29, 72.

Goa S., Tanji K.K., Peters D.W., Herbel M.J. (2000): Water selenium and sediment fractionation in a California flow-through wetland system. Journal of Environmental Quality 29, 1275-1283.

Guelke M. and von Blankenburg F. (2007): Fractionation of stable iron isotopes in higher plants. Environmental Science & Technology 41, 1896–1901.

Guelke-Stelling M. and von Blanckenburg, F. (2012): Fe isotope fractionation caused by translocation of iron during growth of bean and oat as models of strategy I and II plants. Plant and Soil 352, 217–231.

Gunes V., Ozcan K., Citil M., Onmaz A.C., Erdogan H.M. (2010): Detection of myocardial degeneration with point-of-care cardiac troponin assays and histopathology in lambs with white muscle disease. The Veterinary Journal 184 (3), 376-378.

Guo W., Hu S., Wang Y., Zhang L., Hu Z., Zhang J. (2013): Trace determination of selenium in biological samples by CH_4-Ar mixed gas plasma DRC-ICP-MS. Microchemical Journal 108, 106-112.

Hagiwara Y. (2000): Selenium isotope ratios in marine sediments and algae—a reconnaissance study. MS thesis, University of Illinois at Urbana-Champaign, Urbana. IL.

Hasanuzzaman M., Nahar K. Fujita M. (2014): Chapter 16 – Silicon and Selenium: Two Vital Trace Elements that Confer Abiotic Stress Tolerance to Plants. Emerging Technologies and Management of Crop Stress Tolerance, Volume 1, 377-422.

Haug A., Graham R.D., Christophersen O.A., Lyons G.H. (2007): How to use the world's scarce selenium resources efficiently to increase the selenium concentration in food. Microbial Ecology in Health and Disease 19 (4), 209–228.

Herbel M.J., Johnson T.M., Oremland R.S., Bullen T.D. (2000): Fractionation of selenium isotopes during bacterial respiratory reduction of selenium oxyanions. Geochimica et Cosmochimica Acta 64 (21), 3701-3709.

Herbel M.J., Johnson T.M., Tanji K.K., Gao S.D., Bullen T.D. (2002): Selenium Stable Isotope Ratios in California Agricultural Drainage Water Management Systems. Journal of Environmental Quality 31, 1146-1156.

Heuser A. and Eisenhauer A. (2010): A Pilot Study on the Use of Natural Calcium Isotope ($^{44}Ca/^{40}Ca$) Fractionation in Urine as a Proxy for the Human Body Calcium Balance. Bone 46, 889-896.

Heuser A., Eisenhauer A., Gussone N., Bock B., Hansen B.T., Nägler Th.F. (2002): Measurement of calcium isotopes ($\delta^{44}Ca$) using a multicollector TIMS technique. International Journal of Mass Spectrometry 220, 385-397.

Hoefs J. (2009): Stable Isotope Geochemistry. Sixth Edition. Springer, ISBN 9783540707035.

Holben D.H. and Smith A.M. (1999): The diverse role of selenium within selenoproteins: A review. Journal of the American Dietary Association 99 (7), 836–843.

Hultberg B, Andersson A, Isaksson A. (1997): Copper ions differ from other thiol reactive metal ions in their effects on the concentration and redox state of thiols in HeLa cell cultures. Toxicology 117 (2-3), 89-97.

Ip C. (1998): Lessons from basic research in selenium and cancer prevention. Journal of Nutrition 128, 1845-1854.

IPCC (2013): International Panel for Climate Change - Working Group I: Contribution to the IPCC Fifth Assessment Report Climate Change 2013: The Physical Science Basis" (http://www.ipcc.ch/pdf/assessment-report/ar5/wg1/WG1AR5_ALL_FINAL.pdf) (last download 06.12.2015)

Jain J., Gonzales-Gil G., Singh V., Von Hullebusch E.D., Farges F., Lens P.N.L. (2014): Biogenic Selenium Nanoparticles: Production, Characterization and Challenges. In: Kumar (Ed.): Biotechnology Vol. 10: Nano Biotechnology. Studium Press LLC, ISBN 9781626990258.

Johnson T.M., Herbel J.M., Bullen T.D., Zawislanski P.T. (1999): Selenium isotope ratios as indicators of selenium sources and oxyanions reduction. Geochimica et Cosmochimica Acta, Vol.63, No.18, pp.2775-2783.

Johnson T.M., Bullen T.D., Zawislanski P.T. (2000): Selenium Stable Isotope Ratios as Indicators of Sources and Cycling of Selenium: Results from the Northern Reach of San Francisco Bay. Environmental Science & Technology 34, 2075-2079.

Johnson T. (2004): A review of mass-dependent fractionation of selenium isotopes and implications for other heavy stable isotopes. Chemical Geology 204, 201– 214.

Johnson T.M. and Bullen T.D. (2003): Selenium isotope fractionation during reduction of Se oxyanions by Fe(II) + Fe(III) hydroxide-sulfate (green rust). Geochimica et Cosmochimica Acta 67 (3), 413-419.

Jones L., Sever V., Lin Z.Q., Banuelos G.S. (2014): The source-partitioning of selenium volatilization in soil-*Stanleya pinnata* and *Brassica juncea* systems. In: Banuelos, Lin and Yin (Eds.): Selenium in the Environment and Human Health. Taylor and Francis Group, ISBN 9781138000179.

Joy E.J.M, Black C.R., Young S.D., Broadley M.R. (2014): Selenium nutrition in Africa. In: Banuelos, Lin and Yin (Eds.): Selenium in the Environment and Human Health. Taylor and Francis Group 2014, ISBN 9781138000179.

Kahakachchi C., Boakye H.T., Uden P.C., Tyson J.F. (2004): Chromatographic speciation of anionic and neutral selenium compounds in Se-accumulating *Brassica juncea* (Indian mustard) and in selenized yeast. Journal of Chromatography A 1054, 303–312.

Kaneko M. and Poulson S.R. (2012): Rate of Oxygen Isotope Exchange between Selenate and Water. Environmental Science & Technology 46, 4539-4545.

Karadjova I.B., Lampugnani L., Dedina J., D'Ulivo A., Onor M., Tsalev D.L. (2006): Organic solvents as interferents in arsenic determination by hydride generation atomic absorption spectrometry with flame atomization. Spectrochimica Acta Part B 61, 525-531.

Kikkert J. and Berkelaar E. (2013): Plant Uptake and Translocation of Inorganic and Organic Forms of Selenium. Archives of Environmental Contamination and Toxicology 65 (3), 458-465

Kim M.-J. (2001): Separation of Inorganic Arsenic Species in Groundwater using Ion Exchange Method. Bulletin of Environmental Contamination and Toxicology 67, 46-51.

Kirkbright G.F. and Taddia M. (1978): Application of masking agents in minimizing interferences from some metal ions in the determination of arsenic by atomic absorption spectrometry with the hydride generation technique. Analytica Chimica Acta 100, 145-150.

Kopp G. (1999): E900 – Se-Bestimmung. MLS-Applikation Aufschluss. MLS GmbH, Applikationslabor Leutkirch. (available on request).

Krouse H.R. and Thode H.C. (1962): Thermodynamic properties and geochemistry of isotopic compounds of selenium. Canadian Journal of Chemistry 40 (2), 367–375.

Kupka R., Msamanga G.I., Spiegelman D. (2004): Selenium status is associated with accelerated HIV disease progression among HIV 1 infected pregnant woman in Tanzania. Journal of Nutrition 134, 2556-2560.

Layton-Matthews D., Leybourne M.I., Peter J.M., Scott S.D. (2006): Determination of selenium isotopic ratios by continuous-hydride-generation dynamic-reaction-cell inductively coupled plasma-mass spectrometry. Journal of Analytic Atomic Spectrometry 21, 41-49.

Layton-Matthews D., Leybourne M.I., Peter J.M., Scott S.D., Cousens B., Eglington B.M. (2013): Multiple sources of selenium in ancient seafloor hydrothermal systems: Compositional and Se, S, and Pb isotopic evidence from volcanic-hosted and volcanic-sediment-hosted massive sulfide deposits of the Finlayson Lake District, Yukon, Canada. Geochimica et Cosmochimica Acta 117, 313-331.

LeDuc D.L., Tarun A.S., Montes-Bayon M., Meija J., Malit M.F., Wu C.P. (2004): Overexpression of selenocysteine methyltransferase in Arabidopsis and Indian mustard increases selenium tolerance and accumulation. Plant Physiology 135 (1), 377-383.

Legarth J.B. (1996): Sustainable metal resource management—the need for industrial development: efficiency improvement demands on metal resource management to enable a (sustainable) supply until 2050. Journal of Cleaner Production 4 (2), 97-104.

Lenz M. and Lens P.N.L. (2009): The essential toxin: The changing perception of selenium in environmental sciences. Science of the Total Environment 407, 3620-3633.

Lenz M., Floor G.H., Winkel L.H.E., Román-Ross G., Corvini P.F.X. (2012): Online Preconcentration-IC-ICP-MS for Selenium Quantification and Speciation at Ultratraces. Environmental Science & Technology 46, 11988-11994.

Levander O. A. and Burk R. F. (2006): Update of human dietary standards for selenium. In: Hatfield D. L., Berry M. J., Gladyshev V. N. (Eds.): Selenium - Its Molecular Biology and Role in Human Health, 3rd ed. Springer, ISBN 9781461410249.

Li H.-F., McGrath S.P., Zhao F.-J. (2008): Selenium uptake, translocation and speciation in wheat supplied with selenate or selenite. New Phytologist 178, 92-102.

Lin Z.-Q., Cervinka V., Pickering I.J., Zayed A., Terry N. (2002): Managing selenium-contaminated agricultural drainage water by the integrated on-farm drainage management system: the role of selenium volatilization. Water Research 36, 3150-3160.

Longchamp M., Castrec-Rouelle M., Biron P., Bariac T. (2015): Variations in the accumulation, localization and rate of metabolization of selenium in mature Zea mays plants supplied with selenite or selenate. Food Chemistry 182, 128-135.

Mariotti A., Germon J.C., Hubert P., Kaiser P., Letolle R. Tardieux A. Tardieux P. (1981): Experimental determination of nitrogen kinetic isotope fractionation: some principles; illustration for the denitrification and nitrification processes. Plant and Soil 62, 413-430.

Marshall J.R. (2014): Selenium in cancer prevention: Did we move to quickly? In: Banuelos, Lin and Yin (Eds.): Selenium in the Environment and Human Health. Taylor and Francis Group, ISBN 9781138000179.

Miller C.A. (2013): 3.02 Energy Resources and Policy: Vulnerability of Energy Resources and Resource Availability – Fossil Fuels (Oil, Coal, Natural Gas, Oil Shale). Climate Vulnerability – Unterstanding and Addressing Threats to Essential Resources Vol. 3: Vulnerability of Energy to Climate. Elsevier, ISBN 9780123847041, pp. 37-51.

Mitchell K., Mason P.R.D., Van Cappellen P., Johnson T.M., Gill B.C., Owens J.D., Diaz J., Ingall E.D., Reichart G.-J., Lyons T.W. (2012): Selenium as paleo-oceanographic proxy: A first assessment. Geochimica et Cosmochimica Acta 89, 302-317.

Mitchell K., Couture R.M., Johnson T.M., Mason P.R.D., van Cappellen P. (2013): Selenium sorption and isotope fractionation: Iron(III) oxides versus iron(II). Chemical Geology 342, 21-28.

Moreno-Reyes R., Egrise D., Neve J., Pasteels J. L., Schoutens A (2001): Selenium deficiency-induced growth retardation is associated with an impaired bone metabolism and osteopenia. Journal of Bone and Mineral Research 16 (8), 1556–1563.

Moreno-Reyes R., Suetens C., Mathieu F., Begaux F., Zhu D., Rivera M.T., Boelaert M., Neve J., Perlmutter N., Vanderpas J. (1998): Kashin-Beck osteoarthropathy in rural Tibet in relation to selenium and iodine status. The New England Journal of Medicine 339 (16), 1112–1120.

Moss R.L., Tzimas E., Kara H., Willis P., Kooroshy J. (2011): Critical Metals in Strategic Energy Technologies – Assessing Rare Metals as Supply-Chain Bottlenecks in Low-Carbon Energy Technologies. JRC European Commission Scientific and Technical Reports. (https://setis.ec.europa.eu/system/files/CriticalMetalsinStrategicEnergy Technologies-def.pdf) (last download 06.12.2015).

Mounicou S., Vonderheide A.P., Shann J.R., Caruso J.A. (2006): Comparing a selenium accumulator plant (*Brassica juncea*) to a nonaccumulator plant (*Helianthus annuus*) to investigate selenium-containing proteins. Analytical and Bioanalytical Chemistry 386, 1367-1378.

Neal R.H. (1995): Selenium. In: Alloway B.J. (Ed.): Heavy Metals in Soils. Blackie Academic & Professional, London, ISBN 9789400744691, pp.260-283.

Nothstein A. (2015): Selenium Transfer between Kaolinite or Goethite Surfaces, Nutrient Solution and Oryza Sativa. Approved Dissertation, Institute of Applied Geosciences, Karlsruhe Institute of Technology (DOI(KIT): 10.5445/IR/1000049456).

NRM (2015): Natural History Museum (Naturhistoriska riksmuseet), Stockholm, Sweden (http://www.nrm.se/english/researchandcollections/geosciences/vegace nter/instrumentation.9000161.html) (last download 05.12.2015).

Ohlendorf H.M., Hoffman D.J., Salki M.K., Aldrich T.W. (1986): Embryonic mortality and abnormalities of aquatic birds: Apparent impacts of selenium from irrigation drain water. Science of the Total Environment 52, 49–63.

Ohlendorf, H.M. (2002): The birds of Kesterson Reservoir: a historical perspective. Aquatic Toxicology 57, 1-10.

Oldfield J.E. (2002): Selenium World Atlas. Selenium Tellurium Development Association (http://www.369.com.cn/En/Se%20Atlas%202002.pdf) (last download 05.12.2015).

Olesik J.W. and Gray P.J. (2014): Advantages of N_2 and Ar as reaction gases for measurement of multiple Se isotopes using inductively coupled plasma-mass spectrometry with a collision/reaction cell. Spectrochimica Acta Part B 100, 197-210.

Olin A., Noläng B., Öhman L.-O., Osadchii E., Rosén E. (2005): Chemical Thermodynamics Series Volume 7 – Chemical Thermodynamics of Selenium. Thermochemical Database (TDB) Project Publications. Nuclear Energy Agency (NEA), OECD. (https://www.oecd-nea.org/dbtdb/pubs/vol7-selenium.pdf) (last download 05.12.2015).

Pan-pan X., Deng J.-W., Zhang H.-M., Ma Y-H., Cao D.-J., Ma, R.-X., Liu R.-J., Liu C., Liang Y-G. (2015): Effects of cadmium on bioaccumulation and biochemical stress response in rice (*Oryza sativa* L.). Ecotoxicology and Environmental Safety 122, 392-398.

Peak J.D. and Sparks D.L. (2002): Mechanisms of Selenate Adsorption on Iron Oxides and Hydroxides. Environmental Science & Technology 36, 1460-1466.

Penning H. (2005): Fraktionierung stabiler Isotope durch anaerobe Bodenmikroorganismen. Approved dissertation at the Department of Biology, Phillips-University of Marburg/Lahn. (http://archiv.ub.uni-marburg.de/diss/z2005/0526/pdf/dhp.pdf) (last download 05.12.2015).

Pickering I.J., Ponomarenko O., George G.N., La Porte P.F., Strait K., Gailer J., Leslie E.M., Spallholz J. (2014): Synchrotron studies of selenium interactions with arsenic. In: Banuelos, Lin and Yin (Eds.): Selenium in the Environment and Human Health. Taylor and Francis Group, ISBN 9781138000179.

Pillai R., Uyehara-Lock J.H., Bellinger F.P. (2014): Selenium and Selenoprotein Function in Brain Disorders. Critical Review. International Union of Biochemistry and Molecular Biology 66 (4), 229–239.

Pogge von Strandmann P.A.E., Coath C.D., Catling D.C., Poulton A.W., Elliot T. (2014): Analysis of mass dependent and mass independent isotope variability in black shales. Journal of Analytical Atomic Spectrometry 29 (9), 1648-1659.

Pohl P. and Prusisz B. (2004): Ion-exchange column chromatography – an attempt to speciate arsenic. Trends in Analytical Geochemistry 23 (1).

Pokrovsky O.S., Pokrovsky G.S., Schott J., Galy A. (2006): Experimental study of germanium adsorption on goethite and germanium coprecipitation with iron hydroxide: X-ray absorption fine structure and macroscopic characterization. Geochimica et Cosmochimica Acta 70, 3325-3341.

Presser T.S. and Ohlendorf H.M. (1987): Biogeochemical Cycling of Selenium in the San Joaquin Valley, California, USA. Environmental Management 11 (6), 805-821.

Ralston N.V.C. and Raymond L.J. (2014): Selenium status and intake influences mercury exposure risk assesments. In: Banuelos, Lin and Yin (Eds.): Selenium in the Environment and Human Health. Taylor and Francis Group, ISBN 9781138000179.

Rashid K. and Krouse, H.R. (1978): Selenium isotope fractionation during bacterial selenite reduction. In: Zartman, R.E. (Ed.): Short Papers of the Fourth International Conference, Geochronology, Cosmochronology, Isotope Geology. Open-File Report-U.S. Geological Survey, Reston, VA, United States, pp. 347– 348. (http://pubs.er.usgs.gov/publication/ofr78701) (last download 06.12.2015).

Rauland-Rasmusen K., Stupak I., Clarke N. Callesen I., Helmisaari H.-S., Karltun E., Varnagiryte-Kabasinskiene I. (2008): Sustainable Use of Forest Biomass for Energy. In: Bravo F., LeMay V., Jandl R., von Gadow K. (Eds.): Managing Forest Ecosystems – The Challenge of Climate Change (Vol 12). Springer, ISBN 9781402083426, pp. 29-78.

Rayman M.P. (2006): The importance of selenium to human health. Lancet 356, 23-241.

Rees C.B. and Thode H.G. (1966): Selenium isotope effects in the reduction of sodium selenite and of sodium selenate. Canadian Journal of Chemistry 44, 419– 427.

Rees C.E. (1978): Sulphur isotope measurements using SO_2 and SF_6. Geochimica et Cosmochimica Acta 42, 383-389.

Ribeiro A.S., Vieira M.A., Curtius A.J. (2004): Determination of hydride forming elements (As, Sb, Se, Sn) and Hg in environmental reference materials as acid slurries by on-line hydride generation inductively coupled mass spectrometry. Spectrochimica Acta Part B 59, 243-253.

Rouxel O., Fouquet Y., Ludden J.N. (2004): Subsurface processes at the Lucky Strike hydrothermal field, Mid-Atlantic Ridge: Evidence from sulfur, selenium, and iron isotopes. Geochimica et Cosmochimica Acta 68, 2295-2311.

Rouxel O., Ludden J., Carignan J., Marin L., Fouquet Y. (2002): Natural variations of Se isotopic composition determined by hydride generation multiple collector inductively coupled plasma mass spectrometry. Geochimica et Cosmochimica Acta 66 (18), 3191–3199.

Rudge J.F., Reynolds B.C., Bourdon B. (2009): The Double Spike toolbox. Chemical Geology 265, 420-431.

Saito Y., Aramaki K., Hodoshima S., Saito M., Shono A., Kuwano J., Otake K. (2008): Efficient hydrogen generation from organic chemical hydrides by using catalytic reactor on the basis of superheated liquid-film concept. Chemical Engineering Science 63, 4935-4941.

Sandholm M., Oksanen H.E., Pesonen L. (1973): Uptake of selenium by aquatic organisms. Limnology and Oceanography 18, 496–498.

Sarret G., Avoscan L., Carrière M., Collins R., Geoffroy N., Carrot F., Covès J., Gouget B. (2005): Chemical Forms of Selenium in the Metal-Resistant Bacterium Ralstonia metallidurans CH34 Exposed to Selenite and Selenate. Applied Environmental Microbiology 71 (5), 2331–2337.

Savard D., Bédard L.P., Barnes S.-J. (2006): TCF selenium preconcentration in geological materials for determination at sub-µg g^{-1} with INAA (Se/TCF-INAA). Talanta 70, 566-571.

Schilling K., Johnson T.M., Wilcke W. (2011a): Selenium Partitioning and Stable Isotope Ratios in Urban Topsoils. Soil Science Society of America Journal 75, 1354-1364.

Schilling K., Johnson T.M., Wilcke W. (2011b): Isotope Fractionation of Selenium During Fungal Biomethylation by *Alternaria alternate*. Environmental Science & Technology 45, 2670–2676.

Schilling K. and Wilcke W. (2011c): A Method to Quantitatively Trap Volatilized Organoselenides for Stable Selenium Isotope Analysis. Journal of Environmental Quality 40, 1021–1027.

Schilling K., Johnson T.M., Wilcke W. (2013): Isotope fractionation of selenium by biomethylation in microcosm incubations of soil. Chemical Geology 352, 101-107.

Schilling K., Johnson T., Mason P.R.D. (2014): A sequential extraction technique for mass-balanced stable selenium isotope analysis of soil samples. Chemical Geology 381, 125-130.

Schilling K., Johnson T., Dhillon K.S., Mason P.R.D. (2015): Fate of Selenium in Soils at a Seleniferous Site Recorded by High Precision Se Isotope Measurements. Environmental Science & Technology 49 (16), 9690–9698.

Schoenberg R. and von Blanckenburg F. (2005): An assessment of the accuracy of stable Fe isotope ratio measurements on samples with organic and inorganic matrices by high-resolution multicollector ICP-MS. International Journal of Mass Spectrometry 242, 257-272.

Semba R.D., and Gray G.E. (2001): Pathogenesis of anemia during human immunodeficiency virus infection. Journal of Investigative Medicine 49, 225–239.

Sors T.G., Ellis D.R., Salt D.E. (2005): Selenium uptake, translocation, assimilation and metabolic fate in plants. Photosynthesis Research 86 (3), 373-89.

Stone R. (2009): A Medical Mystery in Middle China. Science 324, 1378–1381.

Stranges S., Navas-Acien A., Rayman M.P., Guallar E. (2010): Selenium status and cardiometabolic health: State of the evidence. Nutrition, Metabolism & Cardiovascular Diseases 20, 754–760.

Stueken E., Foriel J., Nelson B. K., Buick R., Catling D.C. (2013): Selenium isotope analysis of organic-rich shales: advances in sample preparation and isobaric interference correction. Journal of Analytical Atomic Spectrometry 28, 1734.

Stueken E., Foriel J., Buick R., Schoepfer S.D. (2015): Selenium isotope ratios, redox changes and biological productivity across the end-Permian mass extinction. Chemical Geology 410, 28-39.

Su C. and Suarez D.L. (2000): Selenate and Selenite Sorption on Iron Oxides: An Infrared and Electrophoretic Study. Soil Science Society of America Journal 64, 101-111.

Sudfeld C.R., Aboud S., Kupka R., Mugusi F.M., Fawzi W.W. (2014): Effect of selenium supplementation on HIV-1 RNA detection in breast milk of Tanzanian women. Nutrition 30, 1081-1084.

Taagepera R. (2014): A world population growth model: Interaction with Earth's carrying capacity and technology in limited space. Technological Forecasting and Social Change 82, 34-41.

Tagmount A., Berken A. and Terry N. (2002): An essential role of S-adenosyl-L-methionine : L-methionine S-methyltransferase in selenium volatilization by plants. Methylation of selenomethionine to selenium-methyl-L-selenium-methionine, the precursor of volatile selenium. Plant Physiology 130, 847–856.

Takeno N. (2005): Atlas of Eh-pH diagrams – Intercomparison of thermodynamic databases. Geological Survey of Japan Open File Report No. 419. (http://www.eosremediation.com/download/Chemistry/Chemical%20Properties/Eh_pH_Diagrams.pdf) (last download 06.12.2015).

Terry N., Carlson C., Raab T.K., Zayed A. (1992): Rates of selenium volatilization among crop species. Journal of Environmental Quality 21, 341-344.

ThermoScientific (2015): Neptune Plus MC-ICP-MS broschure (http://www.thermoscientific.com/content/dam/tfs/ATG/CMD/CMD%20Marketing%20Material/MassSpectrometryCollateral/Brochures/BR30197-NEPTUNE-Plus-Multicollector-ICPMS-EN.pdf) (last download 05.12.2015)

Tsai C.-Y. and Jiang S.-J. (2011): Microwave-assisted Extraction and Ion Chromatography Dynamic Reaction Cell Inductively Coupled Plasma Mass Spectrometry for the Speciation Analysis of Arsenic and Selenium in Cereals. Analytical Sciences 27.

US EPA (2000): Selenium Compounds, Hazard Summary (http://www3.epa.gov/airtoxics/hlthef/selenium.html) (last visit 05.12.2015).

US NRC (2001): US National Research Council. Basic research opportunities in earth science Chapter 4 – The Critical Zone: Earth's Near Surface Environment. National Academy Press, Washington DC, USA. (http://www.nap.edu/read/9981/chapter/4) (last visit 06.12.2015).

USDA (2014): US National Department of Agriculture. Poisonous plant research – selenium accumulating plants. (http://www.ars.usda.gov/Research/docs.htm?docid=9979) (last visit 05.12.2015).

USGS (2015): US Geological Survey pubs.usgs.gov (last visit 22.07.2015).

US-NIH (2013): US National Institute of Health. Selenium – dietary supplement fact sheet (https://ods.od.nih.gov/factsheets/Selenium-HealthProfessional/) (last visit 05.12.2015).

Wachsmann M. and Heumann K.G. (1992): Negative thermal ionization mass spectrometry of main group elements: Part 2. 6th Group: sulfur, selenium, and tellurium. International Journal of Mass Spectrometry and Ion Processes 114, 209– 220.

Wang Z.J. and Gao Y.X. (2001): Biogeochemical cycling of selenium in Chinese environments. Applied Geochemistry 16, 1345–1351.

Wasilewska M., Goessler W., Zischka M., Maichin B., Knapp G. (2002): Efficiency of oxidation in wet digestion procedures and influence from the residual organic carbon content on selected techniques for determination of trace elements. Journal of Analytical Atomic Spectrometry 17, 1121-1125.

Webster C.L. (1972): Selenium isotope analysis and geochemical applications. Approved Dissertation. Colorado State University, Fort Collins, CO.

Weinstein C., Moynier F., Wang K., Paniello R., Foriel J., Catalano J., Pichat S. (2011): Isotope fractionation of Cu in plants. Chemical Geology 286, 266-271.

Welz B. and Melcher M. (1981): Mutual interactions of elements in the hydride technique in atomic absorption spectrometry, Part 1. Influence of Selenium on Arsenic Determination. Analytica Chimica Acta 131, 17-25.

Welz B. and Melcher M. (1984): Mechanisms of Transition Metal Interferences in Hydride Generation Atomic-absorption Spectrometry, Part 1. Influence of Cobalt, Copper, Iron and Nickel on Selenium Determination. Analyst 109, 569-572.

Welz B. (1983): Atomabsorptionsspektrometrie - 3rd edition. Verlag Chemie, ISBN 3527260730, pp. 243 ff.

Wen H., Carignan J., Chu X., Fan H., Cloquet C., Huang J., Zhang Y., Chang H. (2014): Selenium isotope trace anoxic and ferruginous seawater conditions in the Early Cambrium. Chemical Geology 390, 164-172.

Wen H. and Carignan J. (2011): Selenium isotopes trace the source and redox processes in the black shale-hosted Se-rich deposits in China. Geochimica et Cosmochimica Acta 75, 1411-1427.

Wen H. J. and Qiu Y. Z. (2002): Geology and Geochemistry of Se-Bearing Formations in Central China. International Geology Review 44, 164–178.

WHO (2011): World Health Organization. Guideline for Drinking Water Quality. Fourth Edition (http://whqlibdoc.who.int/publications/2011/9789241548151_eng.pdf?ua=1) (last download 05.12.2015)

Wiberg E., Wiberg N., Holleman A.F. (2001): Inorganic Chemistry. Elsevier, ISBN 9780123526519.

Wiederhold J.G. (2015): Metal Stable Isotope Signatures as Tracers in Environmental Geochemistry. Environmental Science & Technology 49 (5), 2606–2624.

Wieser M.E., Holden N., Coplen T.B., Böhlke J.K., Berglund M., Brand W.A., De Bièvre P., Gröning M., Loss R.D., Meija J., Hirata T., Prohaska T., Schoenberg R., O'Conner G., Walczyk T., Yoneda S., Zhu X.-K. (2013): Atomic weights of the elements 2011 (IUPAC Technical Report). Pure and Applied Chemistry 85 (5), 1047-1078.

Winkel L.H.E., Johnson C.A., Lenz M., Grundl T., Leupin O.X., Amini M., Charlet L. (2012): Environmental Selenium Research: From Microscopic

Processes to Global Understanding. Environmental Science & Technology 2012, 46, 571-579.

Wu L. (2004): Review of 15 years of research on ecotoxicology and remediation of land contaminated by agricultural drainage sediment rich in selenium. Ecotoxicology and Environmental Safety 57, 257-69.

Yang B., Ma H-Y., Wang X.-M., Jia Y., Hu J., Li X., Dai C.-C. (2014): Improvement of nitrogen accumulation and metabolism in rice (*Oryza sativa* L.) by the endophyte *Phomopsis liquidambari*. Plant Physiology and Biochemistry 82, 172-182.

Zangrando A.F., Tessone A., Ugan A. Gutierrez M.A. (2014): Applications of Stable Isotope Analysis in Zooarchaeology: An Introduction. International Journal of Osteoarchaeology 24, 127-133.

Zayed A., Lytle C.M., Terry N. (1998): Accumulation and volatilization of different chemical species of selenium by plants. Planta 206, 284–292.

Zeebe R.E. and Wolf-Gladrow D. (2001): CO2 in Seawater: Equilibrium, Kinetics, Isotopes. Elsevier Oceanography Series 65, Amsterdam, ISBN 9780444509468.

Zhang M., Tang S., Huang X., Zhang F., Pang Y. Huang Q., Yi Q. (2014): Selenium uptake, dynamic changes in selenium content and ist influence on photosynthesis and chlorophyll fluorescence in rice (*Oryza sativa* L.). Environmental and Experimental Botany 107, 39-45.

Zhang P. and Sparks D.L. (1990): Kinetics of Selenate and Selenite Adsorption/Desorption at the Goethite/Water Interface. Environmental Science & Technology 24, 1848-1856.

Zhao X. Q., Mitani N., Yamaji N., Shen R.F., Ma J.F. (2010): Involvement of Silicon Influx Transporter OsNIP2;1 in Selenite Uptake in Rice. Plant Physiology 153, 1871-1877.

Zhu J. M., Wang N., Li S., Li L., Su H., Liu C. X. (2008a): Distribution and transport of selenium in Yutangba, China: Impact of human activities. Science of the Total Environment 392, 252–261.

Zhu J.-M., Johnson T., Clark S.K., Zhu X.-K., Wang X.-L. (2014): Selenium redox cycling during weathering of Se-rich shales: A selenium isotope study. Geochimica et Cosmochimica Acta 126, 228-249.

Zhu J.-M., Johnson T.M., Clark S.K., Zhu X.-K. (2008b): High Precision Measurement of Selenium Isotopic Composition by Hydride Generation Multiple Collector Inductively Coupled Plasma Mass Spectrometry with a ^{74}Se-^{77}Se Double Spike. Chinese Journal of Analytical Chemistry 36 (10), 1385-1390.

Zink S., Schoenberg R., Staubwasser M. (2010): Isotopic fractionation and reaction kinetics between Cr(III) and Cr(VI) in aqueous media. Geochimica et Cosmochimica Acta 74, 5729-5745.

Zsolnay, A. (2003): Dissolved organic matter: artefacts, definitions and functions. Geoderma 113, 187-209.

Additional online references for photos and images

www.selenium.de
www.bembu.com
http://drainameducci.blogspot.de
www.upei.ca
www.goatbiology.org
www.salvomessina.com
www.selena.ie
www.ocado.com
www.mhlw.go.jp
www.nature.com

Appendices

Appendix I – list of laboratory equipment and reagents used

Abbreviations

PE – polyethylene
PP – polypropylene
HDPE – high density polyethylene
PTFE – polytetrafluoroethylene
PC – polycarbonate
PFA – perfluoroalkoxy alkanes
p.a. – pro analysi (analytical grade)

Instruments

instrument	model	deliverer	purpose
heating plate (metal free) + controller	T03-312	AHF, Tübingen	heating, evaporation
microwave system	START1500	MLS	plant digestion
vacuum pump	MWO 63/4	KNF Neuberger	vacuum generation
centrifuge	Rotofix 32A	Hettich	centrifugation of TCP
water bath	E30U	Dinkelberg Analytics	continuous heating
hydride generator	FIAS400	Perkin Elmer	purification method (C)
phyto chamber	n/a	York International	plant cultivation
autoclave			sterilization of cultivation equipment
sterile bench	n/a	Böttger	cultivation preparation
plant tissue mill	TissueLyser	Qiagen (Rentsch)	grinding of plant tissue
freeze dryer	Alpha 1-4	Martin Christ	drying of plant tissue

Laboratory equipment

article	material	deliverer	purpose
general			
variable pipettes 0.1-10 µL, 10-100 µL, 100-1000 µL, 500-5000 µL + tips		BioHit mLine, VWR	liquid handling
dash bottle	PFA	AHF	fill up, rinse
flange containers (2 L)	PFA	AHF	cleaning PFA beakers
PFA beakers 7 mL, 17 mL, 22 mL, 30 mL + screw cups	PFA	AHF	sample treatment, storage, evaporation
PFA beakers 30 mL, 60 mL, 90 mL, 500 mL	PFA	AHF	acid and water storage and pipetting
water resistant pens		AHF	PFA labelling
centrifuge tubes 15 mL, 40 mL	PE	VWR	centrifugation, sample treatment
Patho beakers 20, 50, 100 mL	PE	VWR	sample storage
one way syringes and sterile syringe filters (0.45 µm)	PE	VWR	filtration
digestion			
microwave beakers + lids	PFA	MLS	digestion
quartz inlays (30 mL) + lids	quartz	MLS	digestion
phytoagar vacuum filtration			
Büchner funnels 120 mL	ceramic	Roth	phytoagar introduction and filtration
filter flasks	glass	Roth	liquid sample capture
Guko cuffs	rubber	Roth	connection flask-funnel
tubing ½ "	PE	Roth	vacuum introduction
paper filters	cellulose	Roth	phytoagar filtering

purification			
column holder	HDPE	AGW (self-made)	column holding
minicolumns (5 mL) + lids + tip guards	PP	Spectrum Labs	chromatographic purification
minicolumn frits (0.45 µm)	PE	Spectrum Labs	chromatographic purification
wide neck bottles	PE	VWR	cleaning of minicol-umns and frits
mortar + pounder	agate	VWR	TCP grinding
Nalgene bottle 250 mL	HDPE	AHF	TCP production
pH indicator paper (1-14)	paper	VWR	neutralization moni-toring (method (A))
cultivation experiments			
Magenta boxes	PC	Sigma Aldrich	cultivation
Magenta box couplers	PP	Gentaur	box connection
volumetric flasks	glass	VWR	production and quantification of Se doped phytoagar
plant tissue preparation			
Eppendorf cups (safe-lock)	PP	Eppendorf	plant grinding and drying
steel beds (5 mm)	stainless steal	Qiagen (Rentsch)	plant grinding (tissue lyzer)
scissors medium	ceramic	Kyocera	plant minimization
scissors small	stainless steel	SK Stahlwaren	plant minimization
scalpels	stainless steel	Braun	plant part separation

Reagents

reagent	Quality / purity grade	deliverer	purpose
general			
HNO_3	p.a., in-house double distilled	Merck	digestion, purification, cleaning
H_2O_2	suprapure	Merck	digestion, purification (trapping)
HCl	suprapure	Merck	reduction, purification
ethanol	p.a., >99.8 % (denatured)	Roth	cleaning of minicolumns and frits, sterilization of sterile bench + equipment
isopropanol	p.a., >99.5 %	Roth	cleaning of minicolumns and frits
Se isotope analytics			
$^{74}Se/^{77}Se$ Double Spike	>99.8 % purity	Chemotrade	mass bias correction
NIST3149 certified Se isotope standard	n/a	NIST	method calibration, monitoring, validation
digestion			
certified reference material NISTSRM 1567a (Wheat Flour)	n/a	NIST	validation of plant digestion methods, internal Se isotope standard
certified reference material SGR-1 (green river shale)	n/a	USGS	validation of Se isotope analytics
HF	suprapure	Merck	SGR-1 digestion
$HClO_4$	normapure	VWR	SGR-1 digestion

purification			
AG1-X8 anion exchange resin (chloride form, 100-200 µm mesh size)	analytical grade	BioRad	purification (method (A))
cellulose powder (20 µm)	n/a	Sigma Aldrich	production TCP
thioglycolic acid	Pure	AppliChem	production TCP
sulphuric acid	Suprapure	Merck	production TCP
acetic anhydride	Extrapure	Merck	production TCP
acetic acid	Suprapure	Merck	pProduction TCP
$NaBH_4$ granulate	n/a	Merck	HG
NaOH pellets	Pure	Merck	purification method (C) (trapping)
Mucasol detergent	n/a	Sigma Aldrich	cleaning of PFA beakers
cultivation experiments			
ethanol	100 % (not denatured)	VWR/promochem	sterilization of rice seeds
NaClO	p.a.	Merck	sterilization of rice seeds
Sterilium sterilizer	n/a	Hartmann	hand sterilization before sterile bench work
plant tissue preparation			
liquid N_2	technical quality	Air Liquide	freezing (plant tissue preparation)

Appendix II – cleaning procedures

Make sure that cleaning and drying spaces and surfaces are not exposed to Se! Always use millipore water (R = 18.2 Ω) when referred to H_2O!

5 % HNO_3

HNO_3 p.a. (65 %) diluted 1:13 to H_2O

Relevant material characteristics

PFA – tolerates boiling acid up to ~300°C, hydrophobic surface, slightly porous (ensures quantitative recovery of sample Se)

Glass/quartz – tolerates boiling acid, hydrophilic surface, not porous (ensures low blanks)

PE – tolerates diluted HNO_3 only at room temperature, tends to retain HNO_3 (rinse properly!)

Steel – might cause metal contamination (use sparsely and before purification only!)

Cleaning PFA beakers

7, 17, 22, 30 mL – use for samples! (clean after any usage!)

60, 90, 500 mL – use for acid and H_2O storage only! (clean after laboratory session!)

2 L flange containers – use for cleaning procedure only!

Cleaning procedure

1. If solid residuals or scums, scrub with *Mucasol* and Q tip
2. Rinse every beaker and lid individually three times with H_2O
3. Transfer beakers and lids to PFA container
4. Fill container with 5 % HNO_3, close container
5. Put container on hotplate at 200°C, keep for 4 to 7 days
6. Let container cool down for several hours
7. Remove HNO_3, fill container with H_2O, keep several hours
8. Remove H_2O, wash every beaker and lid <u>and the container</u> individually three times with H_2O
9. Transfer beakers and lids to container, fill with H_2O, keep for 1 day
10. Remove H_2O, rinse every beaker and lid individually three times with H_2O
11. Let beakers and lids dry (upside down), close beakers, store in closed and clean facility

Cleaning quartz vessels

Between two digestion batches – clean by using a complete "blank digestion batch" without samples

After digestion session – clean with the same procedure as PFA beakers, wrap into Kimwipes after drying, store in closed and clean facility

Be careful, especially lids are fragile!

Cleaning microwave beakers (PFA)

Between two digestion batches – clean by using a complete "blank digestion batch" without samples

After digestion session – fill with 50 mL 32 % HNO_3 (1:1 conc. HNO_3/H_2O), put on hotplate at 200°C and boil for 2 days, rinse with H_2O, store with microwave equipment

Cleaning PE tweezer, ceramic scissors, scalpels

Clean with ethanol and 5 % HNO_3, rinse with H_2O

Cleaning glass graduated beakers, stir rods and other glass equipment

1. Rinse with H_2O, wipe out
2. Fill beakers with 5 % HNO_3, place smaller equipment into beakers
3. Close beakers with watch glass dish, put on hotplate at 200°C for 4-7 days
4. Remove HNO_3, rinse with H_2O, let beakers dry upside down, store in closed and clean facility

Cleaning magenta boxes (PE)

1. Rinse boxes and couplers with H_2O, wipe out
2. Fill boxes with 5 % HNO_3, place couplers in a large glass container with 5 % HNO_3
3. Cover boxes and glass container with watch glass dishes, keep at room temperature for 4-7 days
4. Remove HNO_3 from boxes, rinse with H_2O, fill boxes with H_2O, keep at room temperature for 1 day
5. Remove H_2O, rinse boxes with H_2O, let boxes dry upside down
6. Remove HNO_3 from glass container, rinse couplers and glass container with H_2O, place couplers into glass container filled with H_2O, keep at room temperature for 2 days, remove H_2O from glass container and rinse with H_2O

Cleaning minicolumns and frits (PE)

1. If used with AG1-X8 resin, let fully dry
2. Remove packing material
3. Remove PE frit: push out of holder with rod or wire (paper clip) and remove from column with compressed air

4. Rinse columns (+cups and tips) with H_2O (at least three times), collect in container
5. Collect frits in a (e.g. 60 mL) PFA beaker, fill with H_2O, shake, remove H_2O, repeat three times
6. Fill both containers and PFA beaker with ethanol (30 %), isopropanole (30 %) and H_2O (1:10) and shake for 30 min
7. Fill both with H_2O, shake, remove, repeat three times (until no more foaming visible)
8. Fill both with 5 % HNO_3, keep at room temperature for 4-7 days, remove HNO_3, fill with H_2O, keep at room temperature for 2 days, remove H_2O, rinse individually with H_2O
9. Remove HNO_3, rinse columns and frits individually, dry in clean environment

Cleaning centrifuge tubes and PE sample containers (one-way) (PE)

Rinse container and lid with H_2O, dry in clean environment

Appendix III – purification procedure instructions

Method (A) – anion exchange

Preparation of packing material – add 50 g AG1-X8 (dry substance) to 250 mL bottle (PFA/glass), wash successively with methanol, 1M NaOH and 1M HCl (add to bottle, shake, settle, remove supernatant), store in 1M HCl

Preparation of samples – evaporate aliquot at 70°C to near dryness, shortly oxidize organic samples with 100 µL conc. HNO_3 + 100 µL conc. H_2O_2, evaporate again to near dryness, dilute to 10 mL H_2O, add 1 µL 0.25mM $K_2S_2O_8$ per 10 ng Se, heat up in closed beakers at 120°C for 90 min for full oxidation

Preparation of columns – fill 1.2 mL AG1-X8 suspension into columns, clean and activate resin by passing 10 mL 6M HCl, pass H_2O until eluate turns pH neutral

Purification – add sample, pass 20 mL H_2O (wash), extract Se with 5 * 1 mL 6M HCl

Method (B) – thiol retention

Preparation of packing material – weight 5 g cellulose powder into 250 mL PFA bottle, add a mixture of 30 mL thioglycolic acid, 15 mL acetic anhydride, 10 mL acetic acid and 0.5 mL sulphuric acid, cap bottle and shake for 30 min, let settle for 2 h, heat up for 24 h at 55°C, cool down, shake for 30 min, heat up for 24 h at 55°C, wash with H_2O, filter and let dry at flying air, mortar to homogeneous powder

Preparation of samples – evaporate aliquot at 70°C to near dryness, shortly oxidize organic samples with 100 µL conc. HNO_3 + 100 µL conc. H_2O_2, evaporate again to near dryness, dilute to 2.5 mL 4M HCl, heat up at 80°C for 90 min to full reduction, dilute to 10 mL 1M HCl

Preparation of columns – fill 0.1 g TCP into columns, condition TCP by passing 2 * 2 mL H_2O, 2 mL 6M HCl, 2 mL 1M HCl (make sure column never dries out)

Purification – add sample, pass 2 mL H_2O, 2 mL 6M HCl, 2 mL 1M HCl (wash), transfer water saturated powder to 50 mL centrifuge tube (e.g. by flushing out with H_2O), centrifuge at 4000 rpm for 10 min, remove supernatant, add 500 µL conc. HNO_3 followed by 500 µL H_2O, heat up at 100°C for 20 min, cool down, add 3 mL H_2O and shake tube, centrifuge at 4000 rpm for 10 min, transfer supernatant to beaker, add 500 µL conc. HNO_3 followed by 500 µL H_2O to residual TCP, heat up at 100°C for 20 min, cool down, add 3 mL H_2O and shake tube, centrifuge at 4000 rpm for 10 min, combine both supernatants (Se extract), remove residual TCP in extract by repeated oxidation with 1 mL 1:10 conc. H_2O_2 and 0.5M HNO_3

Method (C) - hydride separation

Preparation of samples – dilute sample to 4M HCl (if containing HF, evaporate at 70°C first), heat up at 80°C for 90 min to full reduction, dilute to 2M HCl

Hydride generation – introduce sample and $NaBH_4$ solution (2.4 g $NaBH_4$ pellets + 4 g NaOH pellets to 1 L H_2O, pH 10-11) to hydride generator via peristaltic pump (80 RPM), trap gas phase in alkaline peroxide solution (5 mL 1M NaOH + 1 mL 30 % H_2O_2), drain gas-liquid separator, take up 2M HCl for 90 s after full sample uptake

Anion exchange – heat up Se containing alkaline traps at 80°C for 60 min to full oxidation, apply method (A)

(if sample contains high amounts of Fe, Co, Ni, Cu or As, apply anion exchange before hydride generation!)

Appendix IV – raw data

List of abbreviations in sample IDs

pini	phytoagar initial		
pac	phytoagar after cultivation		
cp	cultivated plants		
cpr	cultivated plants - roots		
cps	cultivated plants - shoots		
I to V	MinPaX		
1 to 10	Repetition or box	1	selenate 100 µg L^{-1}
		2	selenate 500 µg L^{-1}
		3	selenate 1000 µg L^{-1}
		4	selenite 100 µg L^{-1}
		5	selenite 500 µg L^{-1}
		6	selenite 1000 µg L^{-1}
		7	SeMet 100 µg L^{-1}
		8	SeMet 500 µg L^{-1}
		9	SeMet 1000 µg L^{-1}
		10	no Se supplied
WF	Wheat Flour NISTSRM1567a reference		
MS	multielement standard		
PP	Punjab plants		
VT	validation test (Se free matrix + NIST-Se)		
cp	cultivated plants		
p	phytoagar		
c	clean and condition		
e	eluate		
w	wash		

Se	Se extract
A, B, C	purification method
S	supernatent (after B - centrifugation)

Table IV-1: Se signal optimization measures for Se isotope analytics with HG-MC-ICP-MS – plasma temperature (~RF power), guard electrode activation and HNO3 vs. HCl as a process reagent

		RF 950W, guard on			RF 1200W, guard off			RF 1200W, guard on			RF 1350W, guard on		
Cup ID	**Set on mass**	**Signal [V]**	**Noise [V]**	**S/N [-]**	**Signal [V]**	**Noise [V]**	**S/N [-]**	**Signal [V]**	**Noise [V]**	**S/N [-]**	**Signal [V]**	**Noise [V]**	**S/N [-]**
L2 - 1	74				0.01228	0.00054		0.03811	0.00181				
L2 - 2	74				0.01252	0.00059		0.03808	0.00170				
L2 - average	74	0.04452	0.00185	24.03996	0.01240	0.00057	21.90724	0.03809	0.00175	21.74315	0.05972	0.00271	22.05318
L1 - 1	76				0.20613	0.08401		0.71376	0.30025				
L1 - 2	76				0.19713	0.07378		0.71482	0.30147				
L1 - average	76	0.76433	0.29692	2.57424	0.20163	0.07890	2.55566	0.71429	0.30086	2.37417	1.04520	0.43763	2.38830
C - 1	77				0.11586	0.00483		0.36581	0.02624				
C - 2	77				0.11974	0.00515		0.35905	0.02624				
C - average	77	0.43920	0.05527	7.94589	0.11780	0.00499	23.60535	0.36243	0.02624	13.81129	0.53180	0.03260	16.31388
H1 - 1	78				0.39051	0.02502		1.12842	0.09955				
H1 - 2	78				0.38255	0.02752		1.12780	0.09682				
H1 - average	78	1.30843	0.09546	13.70701	0.38653	0.02627	14.71152	1.12811	0.09819	11.48929	1.80608	0.14227	12.69466
H2 - 1	82				0.13508	0.00558		0.42308	0.01706				
H2 - 2	82				0.14612	0.00552		0.43086	0.01591				
H2 - average	82	0.48766	0.01712	28.48464	0.14060	0.00555	25.32229	0.42697	0.01648	25.90136	0.69764	0.02506	27.83895

	Reagent	0.2M HCl I			0.2M HCl II			0.2M HNO_3		
Cup ID	Mass	Signal 8 ppb Se standard [V]	Noise (blank) [V]	S/N	Signal 8 ppb Se standard [V]	Noise (blank) [V]	S/N	Signal 8 ppb Se standard [V]	Noise (blank) [V]	S/N
L4	72	0.00020	0.00021	0.95238	0.00018	0.00018	1.01124	0.00020	0.00015	1.33333
L2	74	0.02660	0.00080	33.25000	0.00600	0.00165	3.63636	0.01640	0.00100	16.40000
L1	76	0.47250	0.15440	3.06017	0.00700	0.17700	0.03955	0.35500	0.19320	1.83747
C	77	0.26830	0.01329	20.18811	0.46000	0.01420	32.39437	0.15200	0.01030	14.75728
H1	78	0.83100	0.05611	14.80914	0.86300	0.06000	14.38333	0.49200	0.05090	9.66601
H2	80	25.70600	23.61000	1.08878	25.68000	25.58000	1.00391	29.75000	27.88000	1.06707
H3	82	0.37000	0.01950	18.97436	0.37000	0.02000	18.50000	0.20300	0.01470	13.80952
H4	83	0.00024	0.00022	1.12442	0.00037	0.00030	1.22000	0.00035	0.00042	0.82143

Table IV-2: Isotope abundances and masses of Se, Ge, Ar and Ar hydrides (according to Berglund and Wieser, 2011)

Selenium				Germanium		
Isotope	Mass	Abundance NIST3149	Abundance Double Spike	Isotope	Mass	Abundance
^{74}Se	73.9224767	0.00888927	0.52276057	^{72}Ge	71.9242500	0.2731
^{76}Se	75.9192143	0.09355991	0.00528783	^{73}Ge	72.9234595	0.0776
^{77}Se	76.919148	0.07620204	0.47068008	^{74}Ge	73.9211784	0.3672
^{78}Se	77.9173097	0.23744616	0.0003192	^{76}Ge	75.9214029	0.0783
^{80}Se	79.9165221	0.49669584	0.00089221			
^{82}Se	81.9167003	0.08720677	6.0113E-05			
Ar Dimers				Ar Hydrides		
Isotope	Mass	Abundance		Isotope	Mass	Abundance
$^{40}Ar^{40}Ar$	79.9247662	0.99598915		$^{40}Ar^{40}ArH$	80.9325913	0.99598915
$^{40}Ar^{38}Ar$	77.9251153	0.00063199		$^{40}Ar^{38}ArH$	78.9329404	0.00063199
$^{40}Ar^{36}Ar$	75.9299294	0.00336495		$^{40}Ar^{36}ArH$	76.9377544	0.00336495
$^{38}Ar^{38}Ar$	75.9254644	4.0102E-07		$^{38}Ar^{38}ArH$	76.9332894	4.0102E-07
$^{38}Ar^{36}Ar$	73.9302785	2.1352E-06		$^{38}Ar^{36}ArH$	74.9381035	2.1352E-06
$^{36}Ar^{36}Ar$	71.9350925	1.1369E-05		$^{36}Ar^{36}ArH$	72.9429176	1.1369E-05

Table IV-3: Phytoagar treatment – Se concentration using digestion after Kopp (1999) and vacuum filtration

digestion after Kopp (1999)		Se in digest [µg L^{-1}]					
Se species	Se added to phytoagar [µg L^{-1}]	I	II				average
selenate	100	96.0	59.1				77.5 ±18.5
	500	499	348				424 ±75.6
	1000	891	732				812 ±79.7
selenite	100	55.4	61.3				58.4 ±3.0
	500	410	350				380 ±29.7
	1000	801	639				720 ±81.2
SeMet	100	75.3	77.7				76.5 ±1.2
	500	396	306				351 ±45.1
	1000	815	620				718 ±97.5
vacuum filtration		Se in extract [µg L^{-1}]					
Se species	Se added to phytoagar [µg L^{-1}]	I	II	III	IV	V	average
selenate	100	102	103	103	n/a	n/a	103 ±0.7
	500	522	530	528	n/a	n/a	526 ±3.0
	1000	1070	1080	1080	n/a	n/a	1080 ±1.7
selenite	100	102	103	107	n/a	n/a	103 ±0.5
	500	544	542	547	n/a	n/a	544 ±1.7
	1000	1010	1020	1040	n/a	n/a	1030 ±9.3
SeMet	100	96.4	96.1	97.4	110	107	101 ±5.5
	500	478	482	485	573	525	508 ±32.3
	1000	915	901	896	1140	1030	977 ±88.3

Table IV-4: Initial Se in NISTSRM1567a measured with EDX and Se concentration after using digestion after Bell et al. (1992) compared to Kopp (1999)

	EDX	digestion after Bell et al. (1992)		digestion after Kopp (1999)	
sample No.	**Se in solid [ppm]**	**Se in digest [$\mu g\ L^{-1}$]**	**Se in solid [ppm]**	**Se in digest [$\mu g\ L^{-1}$]**	**Se in solid [ppm]**
1	1.36	9.79	0.98	9.65	0.97
2	1.29	6.12	0.61	9.83	0.98
3	1.39	7.53	0.75	9.99	1.00
4	0.98	7.78	0.78	8.54	0.85
5	1.21	7.86	0.79	9.67	0.97
6		8.48	0.85	9.59	0.96
7		9.28	0.93	10.0	1.00
8		9.35	0.93	7.50	0.75
9		8.87	0.89	11.0	1.10
10		9.02	0.90	11.2	1.12
11		8.04	0.80	11.4	1.14
12		9.02	0.90	11.4	1.14
13		8.44	0.84	10.9	1.09
14		8.60	0.86	10.9	1.09
15		9.58	0.96		
16		9.75	0.98		
17		9.78	0.98		
18		13.6	1.36		
19		9.36	0.94		
20		9.82	0.98		
21		10.8	1.08		
22		8.64	0.86		
23		10.5	1.05		
24		8.98	0.90		
25		10.6	1.06		
26		10.7	1.07		
27		12.4	1.24		
28		10.3	1.03		
29		8.00	0.80		
30		9.14	0.91		
31		9.78	0.98		
average	**1.25 ±0.15**	**9.35 ±1.02**	**0.94 ±0.06**	**10.1 ±0.81**	**1.01 ±0.08**

Table IV-5: Initial organic C content in NISTSRM1567a measured with CSA and TOC residuals in treated phytoagar and plant samples

CSA							
Sample		**C [ppm]**					
NISTSRM1567a – 1		43.3					
NISTSRM1567a – 2		41.4					
NISTSRM1567a – 3		43.9					
NISTSRM1567a – 4		42.6					
NISTSRM1567a – 5		42.6					
average		42.8 ±0.69					
digestion after Kopp (1999) (WF – Wheat Flour NISTSRM1567a)							
	weight [mg]	**TOC [mg L^{-1}]**	**average [mg L^{-1}]**	**average (blank substracted) [mg L^{-1}]**	**$C_{absolute}$ [mg]**	**$C_{initial}$ [mg]**	**$C_{residual}$ [%]**
Blank	0	5.5	5.5 ±0.0	-			
		5.5					
		5.4					
WF1	106	8.0	8.0 ±0.0	2.5 ±0.0	0.03	45.5	0.06
		8.0					
		8.0					
WF2	108	77.4	76.3 ±0.7	70.8 ±0.7	0.71	46.1	1.54
		75.8					
		75.7					
WF3	102	64.1	62.0 ±1.4	56.5 ±1.4	0.57	43.5	1.30
		61.5					
		60.4					
WF4	103	51.2	51.2 ±0.0	45.7 ±0.0	0.46	44.2	1.03
		51.1					
		51.2					
WF5	104	20.1	20.1 ±0.0	14.6 ±0.0	0.15	44.6	0.33
		20.1					
		20.1					
WF6	104	25.2	25.1 ±0.1	19.6 ±0.1	0.20	44.3	0.44
		25.1					
		24.9					
WF7	102	39.5	39.6 ±0.1	34.1 ±0.1	0.34	43.4	0.78
		39.7					
		39.5					
WF8	105	31.1	30.8 ±0.2	25.3 ±0.2	0.25	44.8	0.57
		30.5					
		30.7					
WF9	100	10.9	10.9 ±0.0	5.4 ±0.0	0.05	42.9	0.13
		10.8					
		10.9					
	average		**36.0 ±18.9**	**30.5 ±18.9**	**0.31 ±0.19**	**44.4 ±7.9**	**0.69 ±0.42**

NISTSRM1567a digestion after Bell et al. (1992) (WF – Wheat Flour NISTSRM1567a)						
	sample weight [mg]	TOC [mg L^{-1}]	TOC (blank substracted) [mg L^{-1}]	$C_{initial}$ [wt.%]	$C_{initial}$ [mg]	$C_{residual}$ [%]
Blank	0	8.8	-			
WF1	100	313	304	42.8	42.8	7.11
WF2	100	300	291	42.8	42.8	6.80
WF3	100	351	343	42.8	42.8	8.01
WF4	100	410	402	42.8	42.8	9.39
average		**344 ±37.3**	**334 ±37.3**			**7.83 ±0.87**

Cultivated plant digestion after Kopp et al. (1999) (CP – cultivated plants from MinPaX (complete minimization with Tissue Lyzer))						
	sample weight [mg]	TOC [mg L^{-1}]	TOC (blank substracted) [mg L^{-1}]	$C_{initial}$ [wt.%]	$C_{initial}$ [mg]	$C_{residual}$ [%]
Blank	0	8.8	-			
CP1	1200	5.2	<blank	0.21	2.48	-1.42
CP2	1200	9.0	<blank	0.21	2.48	0.08
CP3	1220	8.1	<blank	0.21	2.53	-0.25
CP4	1110	7.6	<blank	0.21	2.30	-0.52
average		**7.5 ±1.1**	**<blank**			**<0.04**

Cultivated plant digestion after Kopp et al. (1999) (CP – cultivated plants from MinPaX (incomplete minimization))						
	sample weight [mg]	TOC [mg L^{-1}]	TOC (blank substracted) [mg L^{-1}]	$C_{initial}$ [wt.%]	$C_{initial}$ [mg]	$C_{residual}$ [%]
Blank	0	8.8	-			
CPI1	100	88.9	80.1	43.8	43.8	1.83
CPI2	70	205	196	43.8	30.6	6.40
CPI3	100	238	229	43.8	43.8	5.23
CPI4	100	216	208	43.8	43.8	4.74
average			**178 ±49.0**			**4.55 ±1.36**

Residual TOC in phytoagar (P) extracts (vacuum filtration)						
		TOC [mg L^{-1}]	TOC (blank substracted) [mg L^{-1}]			
P1		253	262			
P2		394	403			
P3		304	313			
average			**326 ±51.5**			

Table IV-6: Matrix element and Se concentrations measured in multi-element standard and plants from Punjab for each step of method (A) purification and ubiquitary anions present in samples before addition to column

initial plant digests (applies for (B) as well)															
	Amount taken	Na	Mg	Al	P	Ca	Cr	Fe	Co	Ni	Cu	Zn	Ge	As	Se
Digest	[µL]	[µg L^{-1}]	[µg L^{-1}]	[µg L^{-1}]	[µg L^{-1}]	[µg L^{-1}]	[µg L^{-1}]	[µg L^{-1}]	[µg L^{-1}]	[µg L^{-1}]	[µg L^{-1}]	[µg L^{-1}]	[µg L^{-1}]	[µg L^{-1}]	[µg L^{-1}]
1	675	16600	14600	2000	12600	29900	81.8	14900	7.80	54.8	103	373	3.16	13.0	1480
2	675	77000	64800	681	22400	9130	9.7	605	0.39	7.1	43.4	152	1.02	1.0	1480
3	330	4350	20400	3180	26500	36300	19.3	2320	1.15	16.5	57.5	192	0.93	1.6	3010
4	980	708	28200	1450	11200	31900	29.4	1430	0.77	20.8	33.9	133	0.45	0.9	1020
5	690	5210	9450	4782	18600	43000	23.6	2840	1.58	15.1	26.9	76.6	0.94	1.4	1450
6	945	916	11200	596	13100	60900	13.3	265	0.27	8.2	19.1	68.5	0.47	0.4	1060
7	140	805	58800	3560	20900	46000	24.6	2700	2.85	16.9	37.1	141	0.96	1.8	7270
8	650	12900	1500	3330	11200	28800	75.3	18600	7.51	45.2	82.8	380	6.39	8.7	1540
9	170	3630	2120	1360	20200	32200	32.0	1250	0.54	18.4	58.3	265	0.67	1.4	5850
purification	Na	Mg	Al	P	Ca	Cr	Fe	Co	Ni	Cu	Zn	Ge	As	Se	
Sample	[µg L^{-1}]	[µg L^{-1}]	[µg L^{-1}]	[µg L^{-1}]	[µg L^{-1}]	[µg L^{-1}]	[µg L^{-1}]	[µg L^{-1}]	[µg L^{-1}]	[µg L^{-1}]	[µg L^{-1}]	[µg L^{-1}]	[µg L^{-1}]	[µg L^{-1}]	
AMScI	32500	5.8	7.4	n/a	87.0	1.51	22.9	0.03	0.66	0.4	6.7	0.01	0.09	0.5	
AMScII	26600	5.7	6.2	n/a	69.0	1.63	36.2	0.07	0.73	0.5	5.6	<0.001	0.08	1.1	
AMScIII	30300	6.5	7.6	n/a	68.8	2.34	31.2	0.04	1.42	0.5	11.6	0.07	0.10	0.5	
AMScIV	28900	6.3	8.3	n/a	46.8	1.38	23.6	0.04	0.52	0.5	17.5	<0.001	0.05	0.8	
AMSe1	8290	8100	8050	n/a	8140	6540	7480	7790	7910	7840	6230	58.3	46.3	58.6	
AMSe2	8320	8150	8060	n/a	8510	6620	7220	7940	8110	7950	6870	157	64.7	112	
AMSe3	8400	8450	8340	n/a	8570	6750	7100	8130	8220	8130	7130	240	67.1	144	
AMSe4	8460	8400	8330	n/a	9370	7060	7930	8220	8310	8250	6420	25.1	89.5	33.4	
AMSe5	8020	7830	7780	n/a	8230	6470	7250	7720	7810	7760	5900	71.8	49.2	57.8	
AMSe6	7990	7870	7860	n/a	8960	6310	5940	7740	7840	7740	6520	170	24.2	79.8	
AMSe7	7990	7890	7850	n/a	8460	6300	6600	7620	7720	7600	6050	78.6	33.6	97.0	
AMSe8	8100	7970	7920	n/a	8800	6720	7570	7740	7880	7810	5070	2.85	62.9	42.3	
AMSe9	77.9	5.0	12.0	n/a	61.8	0.15	4.9	<0.01	<0.01	<0.08	3.7	0.20	<0.02	3.2	
AMSeBlank	82.0	4.5	14.1	n/a	59.6	4.95	1530	0.16	1.15	3.5	5.2	2.53	<0.02	1.8	
AMSw1	1850	1730	1800	n/a	2020	1560	2250	1810	1850	1860	3930	0.08	683	340	
AMSw2	1590	1550	1620	n/a	1760	1470	2180	1560	1610	1630	3170	0.07	627	287	
AMSw3	1510	1510	1580	n/a	1630	1440	2350	1480	1520	1540	2900	0.07	726	348	

AMSw4	1610	1590	1650	n/a	1720	1380	1990	1560	1610	1620	3850	0.05	795	402
AMSw5	2100	2110	2180	n/a	2280	1800	2560	2050	2110	2120	4380	0.32	884	305
AMSw6	1730	1720	1780	n/a	1860	1520	2750	1670	1720	1740	3280	0.32	701	252
AMSw7	1870	1890	1990	n/a	2010	1680	2550	1800	1870	1890	3770	0.05	689	252
AMSw8	1990	1940	2010	n/a	2190	1670	2020	1890	1960	1970	5110	0.21	423	192
AMSw9	31.1	3.1	5.3	n/a	93.9	0.23	6.6	<0.01	0.16	0.2	6.5	0.40	<0.02	7.0
AMSwBlank	68.7	4.1	32.4	n/a	79.1	0.36	8.7	<0.01	0.18	0.2	4.7	<0.001	<0.02	0.2
AMSSe1	96.9	6.1	13.6	n/a	87.3	1051	24.3	<0.01	0.60	1.5	5.2	<0.001	2.40	428
AMSSe2	135	9.4	27.3	n/a	104	993	22.4	<0.01	0.60	1.7	4.8	0.03	4.10	377
AMSSe3	116	6.0	21.4	n/a	82.9	959	30.1	<0.01	0.45	1.5	11.7	<0.001	2.40	458
AMSSe4	116	6.4	11.5	n/a	66.0	986	20.9	<0.01	0.30	1.5	4.1	<0.001	2.00	459
AMSSe5	218	7.1	13.4	n/a	89.0	981	29.1	<0.01	0.60	1.8	5.6	<0.001	4.15	427
AMSSe6	113	6.7	18.4	n/a	81.4	1050	241	<0.01	0.45	1.4	4.8	<0.001	14.2	332
AMSSe7	207	11.6	33.1	n/a	263	1700	67.3	0.39	0.80	2.2	<0.1	<0.001	6.00	717
AMSSe8	105	8.2	12.8	n/a	125	803	24.3	<0.01	<0.01	8.4	7.6	<0.001	0.85	413
AMSSe9	137	10.9	16.6	n/a	125	1.05	45.0	0.12	<0.01	1.2	12.1	0.03	<0.02	0.2
AMSSeBlank	83.5	6.6	12.3	n/a	90	0.55	25.3	<0.01	<0.01	<0.08	4.8	<0.001	<0.02	0.2
APPc1	5040	10.4	8.2	41.9	258	0.89	11.0	0.02	0.71	5.0	72.8	0.02	0.59	1.2
APPc2	4790	11.8	8.1	30.5	251	1.06	10.5	0.05	2.48	11.9	143	0.02	0.62	0.3
APPe1	776	478	916	587	1400	4.03	621	0.38	2.81	5.0	45.8	0.33	0.59	19.3
APPe2	330	291	31.5	<4.2	416	0.30	1.5	0.02	0.41	0.6	17.4	<0.001	<0.01	0.2
APPe3	107	453	73.3	550	800	0.74	50.6	0.03	0.50	1.6	7.7	<0.001	0.05	18.3
APPe4	91.4	2140	111	796	2370	2.26	99.5	0.06	1.88	3.0	27.1	<0.001	0.07	42.9
APPe5	218	380	194	520	1650	1.03	106.2	0.06	1.24	1.5	11.6	<0.001	0.07	20.6
APPe6	85.1	845	47.3	203	4320	1.05	11.4	0.02	0.83	1.8	22.4	<0.001	<0.01	18.4
APPe7	24.5	676	45.0	192	4870	4.06	57.7	0.05	0.50	0.9	10.5	<0.001	<0.01	24.2
APPe8	680	760	1560	79.0	1410	1.49	186	0.4	3.27	4.8	30.4	0.64	0.08	0.6
APPe9	66.5	285	10.4	11.0	468	0.30	1.9	0.01	0.46	0.6	36.7	<0.001	<0.01	<0.1
APPeBlank	16.5	2.3	1.5	4.2	49.3	0.26	2.4	<0.01	0.15	0.1	12.7	<0.001	<0.01	<0.1
APPw1	215	194	248	525	422	1.46	307	0.11	0.98	2.0	24.1	0.62	0.62	11.5
APPw2	118	87.1	20.4	1680	179	0.37	35.7	0.01	0.32	2.3	28.9	0.53	0.10	52.9
APPw3	49.0	126	21.0	476	286	0.42	28.9	0.01	0.34	1.2	54.9	<0.001	<0.01	17.3
APPw4	36.1	507	27.5	489	611	0.71	35.8	0.02	0.68	1.0	32.3	<0.001	<0.01	25.9
APPw5	78.8	119	58.4	625	557	0.45	57.7	0.03	0.41	1.6	19.2	<0.001	0.07	10.2

APPw6	44.8	228	14.0	1196	1150	0.40	16.8	0.01	0.44	0.7	29.2	<0.001	<0.01	18.5	
APPw7	23.9	168	11.8	139	1205	0.14	13.7	0.01	0.25	0.3	16.6	<0.001	<0.01	15.8	
APPw8	180	182	436	652	367	1.47	946	0.10	0.77	1.8	42.4	<0.001	0.46	5.0	
APPw9	41.0	78.0	14.8	377	168	0.6	24.6	<0.01	0.35	0.9	11.6	<0.001	0.04	43.0	
APPwBlank	14.6	2.0	4.9	<4.2	52.6	0.12	4.1	<0.01	0.20	0.2	7.2	<0.001	<0.01	<0.1	
APPSe1	14.6	5.5	24.1	<4.2	62.3	0.83	13.8	<0.01	0.29	0.2	4.3	<0.001	0.10	61.3	
APPSe2	18.0	5.4	4.8	<4.2	61.3	0.35	4.6	<0.01	0.22	0.6	15.0	<0.001	0.08	15.3	
APPSe3	35.9	6.1	6.3	13.8	143	0.28	5.5	<0.01	0.40	0.5	40.1	<0.001	<0.01	16.1	
APPSe4	10.7	6.6	5.2	<4.2	58.1	0.41	4.3	<0.01	0.17	0.2	4.6	<0.001	0.08	16.1	
APPSe5	12.0	5.5	10.6	<4.2	81.1	0.34	6.0	<0.01	0.16	0.3	6.6	<0.001	0.01	31.5	
APPSe6	11.4	6.9	4.5	12.38	136	0.59	5.4	<0.01	0.14	0.2	4.0	<0.001	0.11	28.7	
APPSe7	17.0	5.0	5.1	<4.2	134	0.42	5.9	<0.01	0.24	0.2	12.2	<0.001	0.01	72.9	
APPSe8	13.7	5.0	58.9	68.1	73.1	0.91	59.4	<0.01	0.22	0.3	7.6	<0.001	0.06	49.8	
APPSe9	160	7.4	13.9	11.3	144	0.94	6.5	0.02	0.63	0.8	15.5	<0.001	0.14	20.9	
APPSeBlank	125	1.9	2.8	<4.2	45.5	0.37	3.5	<0.01	0.18	0.3	5.5	0.06	0.06	<0.1	

Ubiquitary anions in samples before addition to method (A) column (*milliequivalent absolute in sample (10 mL volumes))								
Anion	**Cl^-**		**NO_3^-**		**PO_4^{3-}**		**SO_4^{2-}**	
Sample	**mg L^{-1}**	**meq***	**mg L^{-1}**	**meq***	**mg L^{-1}**	**meq***	**mg L^{-1}**	**meq***
1	n.a.	-	969	0.16	2.46	0.0008	10.6	0.0027
2	2.60	0.0007	1.88	0.0003	n.a.	-	1.23	0.0003

Table IV-7: Matrix element and Se concentrations measured in multi-element standard and plants from Punjab for each step of method (B) purification

		Mg	Al	P	Ca	Cr	Fe	Co	Ni	Cu	Zn	Ge	As	Se
Sample	[µg L^{-1}]	[µg L^{-1}]	[µg L^{-1}]	[µg L^{-1}]	[µg L^{-1}]	[µg L^{-1}]	[µg L^{-1}]	[µg L^{-1}]	[µg L^{-1}]	[µg L^{-1}]	[µg L^{-1}]	[µg L^{-1}]	[µg L^{-1}]	[µg L^{-1}]
BMScI	38.4	5.9	7.8	n/a	345	0.83	7.3	<0.01	0.56	0.4	16.5	0.13	0.14	1.5
BMScII	73.2	9.0	18.8	n/a	399	0.63	12.1	<0.01	0.77	0.7	18.6	<0.001	0.34	2.6
BMSe3	9850	9660	9500	n/a	9680	9680	9690	9920	9670	4840	9770	0.25	348	19.4
BMSe4	9720	9800	9680	n/a	9640	9580	9590	9680	9510	3820	9580	0.18	570	7.4
BMSeBlank	30.8	5.2	9.2	n/a	141	0.72	16.1	0.04	0.30	0.6	7.9	<0.001	0.23	<0.1
BMSe6	9950	10200	10100	n/a	9800	9680	9660	9840	9630	3670	9730	0.20	1610	7.7
BMSe7	9820	10100	9970	n/a	9640	9590	9600	9730	9610	1210	9700	0.15	1741	33.1
BMSe8	9020	94000	9240	n/a	9140	8800	8790	8970	8870	692	8980	0.22	1080	5.8
BMSe9	9570	9920	9860	n/a	10300	9370	9380	9500	9440	910	9450	0.12	1970	8.6
BMSe10	4890	4650	4580	n/a	4920	4860	4970	4860	4860	2080	4890	0.14	339	12.0
BMSw3	190	188	184	n/a	231	184	192	185	183	4790	192	<0.001	0.35	2.6
BMSw4	243	211	230	n/a	851	199	250	202	198	5490	260	0.08	0.76	2.3
BMSwBlank	72.5	5.1	48.5	n/a	69.2	0.22	10.5	<0.01	0.35	0.8	4.2	<0.001	0.32	<0.1
BMSw6	229	231	247	n/a	307	221	228	224	225	5860	234	0.11	2.45	2.9
BMSw7	267	264	256	n/a	383	255	262	256	260	7860	267	<0.001	4.23	5.8
BMSw8	253	248	250	n/a	400	237	240	237	241	7960	248	0.01	10.2	5.1
BMSw9	412	339	336	n/a	523	326	339	330	336	8280	345	0.01	5.66	4.3
BMSw10	109	99.7	101	n/a	206	99.4	108	98.7	98.4	2740	106	<0.001	0.34	3.5
BMSSe3	67.2	26.8	35.1	n/a	643	3.56	18.1	0.22	1.97	131	12.6	1.03	27.0	681
BMSSe4	93.6	24.6	28.6	n/a	1140	0.37	8.7	0.09	0.49	381	24.5	1.17	53.5	201
BMSSeBlank	131	24.3	27.4	n/a	1410	0.30	7.9	<0.01	0.30	0.8	30.4	0.08	0.51	<0.1
BMSSe6	110	25.0	34.5	n/a	989	1.11	16.0	0.22	1.04	344	32.6	2.62	164	1420
BMSSe7	98.0	22.2	26.9	n/a	827	0.60	13.2	0.12	0.72	959	18.5	7.61	389	2990
BMSSe8	92.5	22.5	34.7	n/a	829	0.66	12.2	0.15	0.69	756	21.8	13.0	569	2210
BMSSe9	106	24.5	30.0	n/a	970	0.80	9.2	0.15	0.72	1010	24.8	28.9	695	4510
BMSSe10	90.4	24.6	60.3	n/a	954	0.46	11.3	0.06	0.44	121	20.7	1.69	34.2	881
BPPc1	212	16.4	14.4	22.3	325	0.33	13.4	0.02	0.77	0.4	11.3	1.38	0.19	1.3
BPPc2	130	7.1	3.8	14.8	152	0.27	4.7	<0.01	0.37	0.4	5.4	0.01	0.26	1.0
BPPe1	1140	963	1250	962	2010	5.79	906	0.55	4.81	7.5	44.2	0.22	0.97	16.8

BPPe2	477	402	47.5	1440	619	1.15	45.9	0.05	0.99	4.0	19.4	0.12	0.10	3.6
BPPe3	112	599	98.0	764	1050	0.77	91.4	0.05	0.74	2.4	11.1	0.35	0.04	1.9
BPPe4	46.0	2320	126	1030	2650	3.12	138	0.09	2.15	3.8	17.8	0.02	0.09	4.2
BPPe5	344	572	276	1270	2570	1.79	205	0.12	1.35	2.5	15.7	0.24	0.11	8.1
BPPe6	107	959	58.8	1130	5010	1.48	29.1	0.03	1.26	2.7	13.9	0.28	0.11	2.0
BPPe7	7.2	716	45.1	250	5240	0.57	45.7	0.05	1.01	1.3	5.9	0.09	0.05	2.5
BPPe8	824	832	1750	724	1580	5.28	1160	0.50	3.22	6.1	35.5	0.46	0.59	3.0
BPPe9	49.8	314	25.7	302	529	0.73	27.2	0.02	0.57	1.4	10.5	0.85	0.03	2.0
BPPeBlank	25.2	3.6	4.4	5.2	97.2	0.23	4.7	<0.01	0.31	0.4	4.6	0.09	0.05	<0.1
BPPw1	16.3	27.3	43.4	26.0	92.3	0.32	25.4	0.04	0.21	0.4	4.2	0.26	0.02	1.0
BPPw2	4.8	12.7	16.0	50.6	45.5	0.14	3.1	<0.01	0.08	0.3	3.4	0.01	0.05	1.4
BPPw3	37.5	21.6	8.1	25.1	143	0.21	8.2	0.02	0.85	5.2	40.4	0.46	0.31	0.6
BPPw4	6.7	71.2	7.5	37.9	106	0.21	7.8	<0.01	0.33	0.3	2.7	<0.001	0.02	0.8
BPPw5	4.3	20.8	10.3	37.1	132	0.49	9.3	<0.01	0.60	0.2	3.7	0.02	0.02	0.5
BPPw6	33.1	31.4	7.5	44.8	270	0.66	7.9	0.02	0.95	3.9	32.4	0.21	0.24	0.8
BPPw7	9.4	23.6	5.9	7.7	191	0.28	5.0	<0.01	0.08	0.3	3.4	0.01	0.07	1.1
BPPw8	9.7	25.3	54.3	23.2	67.7	0.28	36.3	0.02	0.15	0.3	2.8	0.02	0.03	0.8
BPPw9	9.3	10.3	2.9	12.8	38.0	0.27	3.6	<0.01	0.12	0.3	2.5	<0.001	0.02	0.6
BPPwBlank	17.3	3.9	3.9	1.6	95.4	0.18	3.8	<0.01	0.68	2.9	27.5	0.03	0.15	0.1
BS1	9.2	5.4	3.9	3.5	291	0.13	3.4	0.02	0.36	1.0	31.4	<0.001	0.22	1.5
BS2	17.1	6.0	5.1	6.6	363	0.09	3.3	0.06	0.19	0.2	32.3	0.03	0.09	1.5
BPPSe1	23.1	28.0	19.4	45.4	434	0.36	15.2	0.03	0.39	0.7	98.2	0.06	0.16	40.2
BPPSe2	42.2	24.8	17.2	28.6	766	0.23	18.1	<0.01	0.27	0.9	96.2	0.08	0.18	96.2
BPPSe3	26.7	20.5	15.7	45.7	403	0.85	21.4	<0.01	0.25	0.6	63.3	0.02	0.15	77.1
BPPSe4	20.8	18.7	12.4	23.1	334	0.20	14.6	<0.01	0.37	1.0	67.6	0.03	0.26	28.5
BPPSe5	27.4	18.4	18.7	15.6	324	0.24	13.1	<0.01	0.13	0.7	6.1	0.04	0.33	20.8
BPPSe6	29.1	18.3	16.1	21.9	331	0.30	16.7	<0.01	0.16	0.6	8.7	0.05	0.52	50.7
BPPSe7	22.4	21.8	23.0	23.7	314	0.18	50.3	<0.01	1.88	3.4	59.1	0.01	0.25	86.0
BPPSe8	19.7	18.4	13.0	12.4	284	0.37	12.2	<0.01	0.23	0.5	81.3	0.40	0.30	52.8
BPPSe9	29.5	19.7	13.6	28.0	272	1.03	29.8	<0.01	0.45	0.7	75.7	0.02	0.27	94.4
BPPSeBlank	50.4	23.4	15.2	28.0	426	0.34	17.2	<0.01	0.39	0.8	86.6	0.14	0.20	<0.1

Table IV-8: Matrix element and Se concentrations measured in multi-element standard for each step of method (C) purification (Na was not determined due to 1M NaOH disturbances being by far higher than sample Na concentrations)

	Mg	Al	Ca	Cr	Fe	Co	Ni	Cu	Zn	Ge	As	Se
Sample	[µg L^{-1}]	[µg L^{-1}]	[µg L^{-1}]	[µg L^{-1}]	[µg L^{-1}]	[µg L^{-1}]	[µg L^{-1}]	[µg L^{-1}]	[µg L^{-1}]	[µg L^{-1}]	[µg L^{-1}]	[µg L^{-1}]
iMS1	16200	16700	16200	16400	17200	16500	16200	16500	16200	1530	1540	1620
iMS2	16300	16500	16800	16100	16800	16200	15800	16200	16000	1470	1520	1510
HGMS1	8.0	20.5	288	9.26	39.6	<0.006	38.9	<0.1	10.8	<0.006	<0.02	<0.1
HGMS2	10.9	27.5	378	9.00	41.0	<0.006	39.1	<0.1	13.3	<0.006	<0.02	<0.1
AEMSe1	10.7	37.5	370	1.80	47.6	<0.006	<0.08	<0.1	15.9	<0.006	<0.02	<0.1
AEMSe2	10.7	29.0	365	0.48	42.8	<0.006	<0.08	<0.1	19.6	<0.006	<0.02	<0.1
AEMSw1	7.1	9.0	110	0.18	10.7	0.02	0.4	0.4	3.8	0.02	0.20	0.4
AEMSw2	3.6	6.2	77.8	0.20	8.8	0.03	0.4	0.3	2.9	0.03	0.06	0.3
AEMSSe1	3.2	3.1	58.7	11.4	1.8	0.12	37.9	0.2	1.7	<0.006	0.26	0.5
AEMSSe2	2.7	2.7	31.3	11.6	2.1	0.10	38.2	0.2	1.6	<0.006	0.26	0.4
AEMSSeBlank	5.3	6.7	177	42.7	42.7	17.0	2.8	0.2	66.0	0.48	2.28	<0.1

Table IV-9: Se concentrations measured in purified samples of methods (A) (including modifications), (B) and (C) for all matrices used (values for multi-element standard and Punjab Plants included in Tables IV-6 to IV-8)

Modifications (performed with Punjab Plants (PP))				
Se [µg L^{-1}]	Ellis et al. (2003)	Variation I	Variation II	Variation III
PP1	118	43.8	66.5	92.2
PP2	30.0	21.3	208	96.6
PP3	31.2	83.1	24.5	99.8
PP4	31.0	19.2	68.6	134
Average (n=4)	**52.5 ±32.6**	**41.9 ±21.6**	**91.8 ±58.0**	**106 ±14.0**
Se only (initial 1000 ng Se)				
Se [µg L^{-1}]	**(A)**	**(B)**	**(C) - after HG**	**(C) - after HG+AE**
Se_1	199	107	132	n/a
Se_2	200	105	115	n/a
Se_3	203	109		
Average	**50.1 ±0.4**	**107 ±1.6**	**124 ±8.3**	
Se free cultivated plants (cp) + doped Se (1000 ng)				
Se [µg L^{-1}]	**(A)**	**(B)**	**(C) - after HG**	**(C) - after HG+AE**
Se_cp_1	193	114	138	21.6
Se_cp_2	199	111	133	24.9
Se_cp_3	203	110	166	50.4
Se_cp_4			166	56.4
Se_cp_5			125	118
Se_cp_6			156	152
Se_cp_7			131	39.3
Se_cp_8			137	70.8
Se_cp_9			139	106
Se_cp_10			143	108
Average	**198 ±2.8**	**111 ±1.5**	**143 ±11.7**	**74.8 ±37.2**
Se free phytoagar (p) + doped Se (1000 ng)				
Se [µg L^{-1}]	**(A)**	**(B)**	**(C) - after HG**	**(C) - after HG+AE**
Se_p_1	131	114	159	94.4
Se_p_2	115	116	149	86.7
Se_p_3	48.4	114		
average	**98.2 ±24.9**	**115 ±1.2**	**154 ±4.9**	**90.5 ±3.9**
Reference materials used for validation (chapter 4.5.3) (WF-Wheat Flour NISTSRM1567a)				
Se [µg L^{-1}]	**(A)**	**(B)**	**(C) - after HG**	**(C) - after HG+AE**
WF1	n/a	n/a	161	114
WF2	n/a	n/a	165	96.1
SGR-1	n/a	n/a	n/a	45.1

Table IV-10: TOC residuals measured in purified plant and phytoagar samples of methods (A), (B) and (C)

	TOC in plant digests [mg L^{-1}]			**TOC in phytoagar extracts [mg L^{-1}]**		
purification method	**A**	**B**	**C**	**A**	**B**	**C**
I	10.5	12.9	<0.9	52.8	57.3	<0.9
II	7.4	20.1	<0.9	121	80.5	<0.9
III	5.6	26.8	<0.9	45.2	11.8	<0.9
average	**7.8 ±1.8**	**20.0 ±4.7**	**<0.9**	**73.0 ±32.0**	**49.9 ±25.4**	**<0.9**

Table IV-11: Uncorrected and corrected Se isotope ratios, internal errors, instrumental mass bias and Se recoveries determined in internal reproducibility and validation test samples of methods (A), (B) and (C) (correction = NIST subtraction) (*added before digestion **added after digestion, 1) after HG and anion exchange

Purification method	Sample ID	matrix	DS addition	$\delta^{82/76}Se$ [‰] (2SD filter)	internal error (2 SE)	$\delta^{82}Se$ [‰]	β_{instr}	Se recovery after HG [%]	total Se recovery [%][1)]
Internal Reproducibility									
(B)	Pl_I1 B	cultivated plant	**	-1.91	0.08	-1.37	-2.13	n/a	n/a
	Pl_I1 B	cultivated plant	**	-1.91	0.09	-1.37	-2.13	n/a	n/a
	Pl_I2 B	cultivated plant	**	-1.86	0.08	-1.33	-2.14	n/a	n/a
	Pl_I2 B	cultivated plant	**	-1.85	0.08	-1.33	-2.13	n/a	n/a
	Ag_I1 B	phytoagar	**	-2.06	0.05	-1.53	-2.12	n/a	n/a
	Ag_I1 B	phytoagar	**	-2.05	0.04	-1.52	-2.12	n/a	n/a
	Ag_I2 B	phytoagar	**	-1.27	0.05	-0.74	-2.12	n/a	n/a
	Ag_I2 B	phytoagar	**	-1.25	0.04	-0.73	-2.13	n/a	n/a
Validation									
(A)	Pl1 A	Se doped plant	**	-0.77	0.03	0.42	-1.98	n/a	96.3
	Pl2 A	Se doped plant	**	2.42	0.06	3.58	-1.90	n/a	99.5
	Ag1 A	Se doped phytoagar	**	5.57	0.02	6.74	-1.85	n/a	65.5
	Ag2 A	Se doped phytoagar	**	3.14	0.02	4.32	-1.88	n/a	57.7
(B)	Pl1 B	Se doped plant	*	24.71	0.07	26.11	-1.71	n/a	5.2
	Pl2 B	Se doped plant	*	11.42	0.12	12.82	-1.90	n/a	2.4
	Pl3 B	Se doped plant	*	21.08	0.12	22.48	-1.76	n/a	2.6
	Pl4 B	Se doped plant	*	31.53	0.09	32.94	-1.59	n/a	12.3
	Pl5 B	Se doped plant	*	34.58	0.20	35.98	-1.55	n/a	1.1
	Pl6 B	Se doped plant	*	25.40	0.82	26.80	-1.70	n/a	0.3
	Pl7 B	Se doped plant	**	-0.57	0.16	0.86	-1.88	n/a	1.7
	Pl8 B	Se doped plant	**	-4.65	0.19	-3.22	-1.92	n/a	1.3
	Pl9 B	Se doped plant	**	4.99	0.05	6.40	-1.80	n/a	25.5
	Pl10 B	Se doped plant	**	33.81	0.21	35.22	-1.45	n/a	4.9
	Pl11 B	Se doped plant	**	17.91	0.04	19.31	-1.74	n/a	29.4
	Pl12 B	Se doped plant	**	13.58	0.20	14.98	-1.83	n/a	5.5
	Ag1 B	Se doped phytoagar	**	19.89	0.20	21.29	-1.68	n/a	4.2
	Ag2 B	Se doped phytoagar	**	35.67	0.32	37.07	-1.46	n/a	2.6

(c)	Pl1 C	Se doped plant	*	-0.14	0.16	0.74	-2.01	83.0	13.0
	Pl2 C	Se doped plant	*	1.21	0.34	2.10	-1.95	79.5	14.9
	Pl3 C	Se doped plant	**	-1.00	0.03	0.00	-2.04	79.5	14.9
	Pl3 C	Se doped plant	**	-1.00	0.02	0.00	-2.05	99.5	30.2
	Pl4 C	Se doped plant	**	-0.98	0.02	0.02	-2.07	99.9	33.9
	Pl5 C	Se doped plant	**	-0.44	0.05	0.40	-1.99	75.2	71.0
	Pl6 C	Se doped plant	**	-1.09	0.05	-0.24	-2.02	93.7	91.3
	Pl7 C	Se doped plant	**	-3.65	0.04	0.37	-2.02	78.5	23.6
	Pl8 C	Se doped plant	**	-3.71	0.03	0.34	-2.02	81.9	42.5
	Pl9 C	Se doped plant	**	-3.71	0.04	0.34	-2.04	83.5	63.8
	Pl10 C	Se doped plant	**	-3.70	0.05	0.33	-2.04	85.6	64.9
	Ag1 C	Se doped phytoagar	**	-3.25	0.04	1.04	-2.01	95.1	56.6
	Ag1 C	Se doped phytoagar	**	-3.25	0.05	1.04	-2.00	95.1	56.6
	Ag2 C	Se doped phytoagar	**	-3.27	0.07	1.13	-2.01	89.3	52.0
	SGR-1	SGR-1	**	-3.62	0.10	0.74	-2.01	n/a	27.0
	WF1	NISTSRM1567a	**	-3.99	0.07	0.34	-1.99	96.4	68.3
	WF2	NISTSRM1567a	**	-4.32	0.05	0.19	-2.00	99.1	57.6

Table IV-12: Se species determined in a plant digest (according to Bell et al. 1992), in the oxidized and reduced samples added to columns of methods (A) and (B) and their respective Se extracts as well as in phytoagar after cultivation (n.q.: not quantifiable though visible peak; - : no visible peak)

		Ion Exchange (Bird et al. 1997)			Ion Pair (Bird et al. 1997)		
Sample	**Details**	**selenate**	**selenite**	**unknown**	**Inorg.Se**	**Org. Se**	**Share organic**
		[ppb]	**[ppb]**	**[ppb]**	[ppb]	[ppb]	[%]
PP7	**Plant digest (Bell et al., 1992)**	311	116	427	183	34.1	15.7
AO1	**Oxidized with K2S2O8**	150	-	150	n/a	n/a	n/a
AO4		131	-	131	n/a	n/a	n/a
BR1	**Reduced with 4M HCl**	-	63.7	<0.1	n/a	n/a	n/a
BR4		-	27.7	<0.1	n/a	n/a	n/a
APPSe1	**Method (A) Se extract**	68.1	-	68.1	n/a	n/a	n/a
APPSe4		104	-	104.4	n/a	n/a	n/a
BPPSe1	**Method (B) Se extract**	-	135	134.7	n/a	n/a	n/a
BPPSe4		-	67.3	67.3	n/a	n/a	n/a
Sample	**Species added**	**selenate**	**selenite**	**unknown**	**selenite**	**SeMet**	**total**
		[ppb]	**[ppb]**	**[ppb]**	**[ppb]**	**[ppb]**	**[ppb]**
IVpac1	**selenate**	n.q.	-	-	-	-	13.5
IVpac2	**selenate**	87.4	-	4.8	-	-	101
IVpac3	**selenate**	889	-	-	-	-	775
IVpac4	**selenite**	-	3.9	2.9	n.q.	-	36.4
IVpac5	**selenite**	-	85.9	12.4	70.1	-	216
IVpac6	**selenite**	-	104	-	105	-	190
IVpac7	**SeMet**	-	-	3.4	31.8	6.6	26.6
IVpac8	**SeMet**	-	-	12.0	n.q.	n.q.	94.5
IVpac9	**SeMet**	-	-	60.8	-	60.1	212
Vpac1	**selenate**	n.q.	-	-	-	-	8.8
Vpac2	**selenate**	136	-	-	-	-	154
Vpac3	**selenate**	743	-	-	-	-	679
Vpac4	**selenite**	-	15.1	-	2.2	23.3	33.4
Vpac5	**selenite**	-	14.9	14.3	n.q.	-	46.0
Vpac6	**selenite**	-	68.6	-	77.0	-	134
Vpac7	**SeMet**	-	-	2.9	1.9	0.7	21.9
Vpac8	**SeMet**	-	-	22.5	n.q.	-	82.4
Vpac9	**SeMet**	-	-	45.7	17.1	4.1	168

Table IV-13: Determination of plant growth during cultivation time for MinPaX I and II (uncertaincy ~0.1 cm) (data on single plant survey within boxes available on request)

MinPaX I	**Average length of plants [cm] (16 = harvest)**					
days from seed	**5**	**7**	**9**	**12**	**14**	**16**
selenate100	0.5	0.6	1.1	4.5	4.2	14.3
selenate500	0.4	0.9	2.3	4.2	4.8	11.4
selenate1000	0.5	1.0	1.6	3.0	2.3	6.3
selenite100	0.4	1.2	3.1	7.4	9.3	15.6
selenite500	0.4	1.3	2.4	5.7	5.6	10.8
selenite1000	0.4	1.1	2.2	3.5	3.4	10.4
SeMet100	0.5	1.3	3.2	6.3	7.6	13.9
SeMet500	0.3	0.7	1.7	4.5	5.2	11.6
SeMet1000	0.4	1.0	1.7	3.4	3.7	9.5
Blank box	0.4	0.7	1.9	4.9	6.4	14.6

MinPaX II						
days from seed	**5**	**7**	**10**	**12**	**16**	
selenate100	0.8	1.5	2.3	3.3	6.7	
selenate500	0.9	2.2	5.2	7.9	12.1	
selenate1000	0.9	1.9	3.9	6.2	10.1	
selenite100	0.9	2.3	5.0	7.2	10.8	
selenite500	0.9	2.0	4.4	6.6	10.3	
selenite1000	0.3	1.3	2.7	4.0	9.4	
SeMet100	0.9	2.1	4.7	7.8	10.0	
SeMet500	0.7	1.5	3.3	5.3	11.2	
SeMet1000	0.8	1.7	3.3	5.1	8.6	
Blank box	0.8	1.9	5.2	8.8	12.8	

Table IV-14: Plant heights and masses after cultivation of MinPaX I-V (uncertaincy length <0.1 cm, phytomass <0.001)

	MinPaX I		MinPaX II		MinPaX III		MinPaX IV		MinPaX V	
mass = DW	**lenght [cm]**	**phytomass [g]**	**lenght [cm]**	**phytomass [g]**	**lenght [cm]**	**phytomass [g]**	**lenght [cm]**	**phytomass [g]**	**lenght [cm]**	**phytomass [g]**
selenate100	14.3	0.09	6.7	0.14	11.6	0.16	9.4	0.11	11.0	0.14
selenate500	11.4	0.10	12.1	0.22	8.6	0.14	6.0	0.07	5.9	0.08
selenate1000	6.3	0.07	10.1	0.23	5.6	0.11	3.4	0.06	4.8	0.10
selenite100	15.6	0.14	10.8	0.19	11.8	0.17	11.1	0.09	5.1	0.03
selenite500	10.8	0.18	10.3	0.18	7.8	0.14	9.9	0.12	8.8	0.12
selenite1000	10.4	0.10	9.4	0.14	9.2	0.16	6.1	0.07	6.0	0.08
SeMet100	13.9	0.12	10.0	0.21	12.3	0.19	8.8	0.11	11.5	0.15
SeMet500	11.6	0.13	11.2	0.16	9.5	0.17	7.8	0.11	4.7	0.04
SeMet1000	9.5	0.10	8.6	0.23	10.4	0.18	6.9	0.12	6.7	0.12
no Se	14.6	0.11	12.8	0.23	10.8	0.19	11.5	0.14	12.0	0.12

Table IV-15: Se concentrations measured in plants (roots + shoots) and phytoagar after cultivation of MinPaX I-V (dig – plant digest, fil – phytoagar filtrate)

MinPaX I				MinPaX V			
Sample ID	**weight sample [mg]**	**Se in dig/fil [µg L^{-1}]**	**Se in solid [ppm]**	**Sample ID**	**weight sample [mg]**	**Se in dig/fil [µg L^{-1}]**	**Se in solid [ppm]**
Ipac1		2.7		**Vpac1**		3.4	
Ipac2		15.9		**Vpac2**		194	
Ipac3		78.1		**Vpac3**		954	
Ipac4		4.9		**Vpac4**		93.8	
Ipac5		48.5		**Vpac5**		418	
Ipac6		113		**Vpac6**		966	
Ipac7		2.5		**Vpac7**		26.1	
Ipac8		25.2		**Vpac8**		310	
Ipac9		61.0		**Vpac9**		413	
Ipac10		<0.1		**Vpac10**		<0.1	
Icpr1	49	125	25.6	**Vcpr1**	55.9	28.7	1.60
Icpr2	52	619	119	**Vcpr2**	27.2	188	5.12
Icpr3	39	1370	357	**Vcpr3**	36.0	197	7.08
Icpr4	74	268	36.0	**Vcpr4**	13.3	32.9	0.44
Icpr5	125	1150	92.1	**Vcpr5**	51.3	128	6.56
Icpr6	62	857	139	**Vcpr6**	37.0	128	4.72
Icpr7	64	312	48.9	**Vcpr7**	63.1	52	3.26
Icpr8	87	1440	167	**Vcpr8**	24.0	381	9.14
Icpr9	63	2390	381	**Vcpr9**	44.4	734	32.6
Icpr10	72	213	29.8	**Vcpr10**	48.6	0.3	0.02
Icps1	40	488	121	**Vcps1**	80.2	76.9	6.17
Icps2	43	2460	573	**Vcps2**	55.1	289	15.9
Icps3	29	1610	561	**Vcps3**	60.9	37.5	2.28
Icps4	67	153	23.0	**Vcps4**	14.7	14.3	0.21
Icps5	59	304	51.7	**Vcps5**	66.2	43.2	2.86
Icps6	39	248	64.2	**Vcps6**	40.7	63.3	2.58
Icps7	60	255	42.9	**Vcps7**	85.2	52.2	4.45
Icps8	46	839	183	**Vcps8**	18.5	111	2.06
Icps9	42	1720	411	**Vcps9**	75.8	224	17.0
Icps10	42	6.9	1.70	**Vcps10**	76.2	0.3	0.02

MinPaX II				MinPaX III				MinPaX IV			
Sample ID	**weight sample [mg]**	**Se in digest [µg L-1]**	**Se in solid [ppm]**	**Sample ID**	**weight sample [mg]**	**Se in digest [µg L-1]**	**Se in solid [ppm]**	**Sample ID**	**weight sample [mg]**	**Se in extract [µg L-1]**	**Se in solid [ppm]**
IIpac1		95.1		**IIIpac1**		0.6		**IVpac1**		3.9	
IIpac2		5.1		**IIIpac2**		9.5		**IVpac2**		123	
IIpac3		3.4		**IIIpac3**		121		**IVpac3**		1000	
IIpac4		0.9		**IIIpac4**		2.6		**IVpac4**		59.0	
IIpac5		49.4		**IIIpac5**		90.3		**IVpac5**		364	
IIpac6		124		**IIIpac6**		45.6		**IVpac6**		977	
IIpac7		9.2		**IIIpac7**		2.2		**IVpac7**		26.9	
IIpac8		30.8		**IIIpac8**		17.7		**IVpac8**		145	
IIpac9		51.6		**IIIpac9**		30.5		**IVpac9**		375	
IIpac10		0.5		**IIIpac10**		0.9		**IVpac10**		<0.1	
IIcp1	71	2490	351	**IIIcp1**	106	404	38.1	**IVcp1**	110	72.8	8.01
IIcp2	97	268	27.6	**IIIcp2**	101	2230	221	**IVcp2**	72	445	32.0
IIcp3	98	2340	239	**IIIcp3**	81	1370	169	**IVcp3**	58	162	9.39
IIcp4	89	427	48.3	**IIIcp4**	105	403	38.2	**IVcp4**	90	43.7	3.93
IIcp5	88	821	93.8	**IIIcp5**	75	804	108	**IVcp5**	115	123	14.2
IIcp6	71	770	108	**IIIcp6**	95	1400	147	**IVcp6**	75	127	9.53
IIcp7	96	154	16.0	**IIIcp7**	83	285	34.3	**IVcp7**	107	57.0	6.10
IIcp8	70	1490	212	**IIIcp8**	88	1640	185	**IVcp8**	115	282	32.3
IIcp9	99	3170	321	**IIIcp9**	78	2620	336	**IVcp9**	116	403	46.7
IIcp10	104	3.0	0.30	**IIIcp10**	77	2.9	0.37	**IVcp10**	143	1.7	0.24

In der Reihe „Karlsruher Geochemische Hefte"
(ISSN 0943-8599) sind erschienen:

Band 1: U. Kramar (1993)
Methoden zur Interpretation von Daten der geochemischen Bachsedimentprospektion am Beispiel der Sierra de San Carlos/ Tamaulipas Mexiko

Band 2: Z. Berner (1993)
S-Isotopengeochemie in der KTB - Vorbohrung und Beziehungen zu den Spuren-elementmustern der Pyrite

Band 3: J.-D. Eckhardt (1993)
Geochemische Untersuchungen an jungen Sedimenten von der Galapagos-Mikroplatte: Hydrothermale und stratigraphisch signifikante Signale

Band 4: B. Bergfeldt (1994)
Lösungs- und Austauschprozesse in der ungesättigten Bodenwasserzone und Auswirkungen auf das Grundwasser

Band 5: M. Hodel (1994)
Untersuchungen zur Festlegung und Mobilisierung der Elemente As, Cd, Ni und Pb an ausgewählten Festphasen unter besonderer Berücksichtigung des Einflusses von Huminstoffen.

Band 6: T. Bergfeldt (1995)
Untersuchungen der Arsen- und Schwermetallmobilität in Bergbauhalden und kontaminierten Böden im Gebiet des Mittleren Schwarzwaldes

Band 7: M. Manz (1995)
Umweltbelastungen durch Arsen und Schwermetalle in Böden, Halden, Pflanzen und Schlacken ehemaliger Bergbaugebiete des Mittleren und Südlichen Schwarzwaldes.

Band 8: J. Ritter (1995)
Genese der Mineralisation Herrmanngang im Albtalgranit (SE-Schwarzwald) und Wechselwirkungen mit dem Nebengestein.

Band 9: J. Castro (1995)
Umweltauswirkungen des Bergbaus im semiariden Gebiet von Santa Maria de la Paz, Mexiko.

Band 10: T. Rüde (1996)
Beiträge zur Geochemie des Arsens.

Band 11: J. Schäfer (1998)
Einträge und Kontaminationspfade Kfz-emittierter Platin-Gruppen-Elemente (PGE) in verschiedenen Umweltkompartimenten.

Band 12: M. A. Leosson (1999)
A Contribution to the Sulphur Isotope Geochemistry of the Upper Continental Crust: The KTB Main Hole - A Case Study

Band 13: B. A. Kappes (2000)
Mobilisierbarkeit von Schwermetallen und Arsen durch saure Grubenabwässer in Bergbaualtlasten der Ag-Pb-Zn-Lagerstätte in Wiesloch

Band 14: H. Philipp (2000)
The behaviour of platinum-group elements in petrogenetic processes: A case study from the seaward-dipping reflector sequence (SDRS), Southeast Greenland volcanic rifted margin

Band 15: E. Walpersdorf (2000)
Nähr- und Spurenelementdynamik im Sediment/Wasser-Kontaktbereich nach einer Seekreideaufspülung – Pilotstudie Arendsee –

Band 16: R. H. Kölbl (2000)
Models of hydrothermal plumes by submarine diffuse venting in a coastal area: A case study for Milos, South Aegean Volcanic Arc, Greece

Band 17: U. Heiser (2000)
Calcium-rich Rhodochrosite layers in Sediments of the Gotland Deep, Baltic Sea, as Indicators for Seawater Inflow

In der Fortsetzungsreihe „Karlsruher Mineralogische und Geochemische Hefte" (ISSN 1618-2677) sind bisher erschienen:

Band 18: S. Norra (2001)
Umweltgeochemische Signale urbaner Systeme am Beispiel von Böden, Pflanzen, und Stäuben in Karlsruhe

Band 19: M. von Wagner, Alexander Salichow, Doris Stüben (2001)
Geochemische Reinigung kleiner Fließgewässer mit Mangankiesen, einem Abfallsprodukt aus Wasserwerken (GReiFMan)

Band 20: U. Berg (2002)
Die Kalzitapplikation als Restaurierungsmaßnahme für eutrophe Seen – ihre Optimierung und Bewertung

Band 21: Ch. Menzel (2002)
Bestimmung und Verteilung aquatischer PGE-Spezies in urbanen Systemen

Band 22: P. Graf (2002)
Meta-Kaolinit als latent-hydraulische Komponente in Kalkmörtel

Band 23: D. Buqezi-Ahmeti (2003)
Die Fluidgehalte der Mantel-Xenolithe des Alkaligesteins-Komplexes der Persani-Berge, Ostkarpaten, Rumänien: Untersuchungen an Fluid-Einschlüssen

Band 24: B. Scheibner (2003)
Das geochemische Verhalten der Platingruppenelemente bei der Entstehung und Differenzierung der Magmen der Kerguelen-Flutbasaltprovinz (Indischer Ozean)

Ab Band 25 erscheinen die Karlsruher Mineralogischen und Geochemischen Hefte bei KIT Scientific Publishing online unter der Internetadresse http://www.ksp.kit.edu
Auf Wunsch sind bei KIT Scientific Publishing gedruckte Exemplare erhältlich („print on demand").

Band 25: A. Stögbauer (2005)
Schwefelisotopenfraktionierung in abwasserbelasteten Sedimenten - Biogeochemische Umsetzungen und deren Auswirkung auf den Schwermetallhaushalt

Band 26: X. Xie (2005)
Assessment of an ultramicroelectrode array (UMEA) sensor for the determination of trace concentrations of heavy metals in water

Band 27: F. Friedrich (2005)
Spectroscopic investigations of delaminated and intercalated phyllosilicates

Band 28: L. Niemann (2005)
Die Reaktionskinetik des Gipsabbindens: Makroskopische Reaktionsraten und Mechanismen in molekularem Maßstab

Band 29: F. Wagner (2005)
Prozessverständnis einer Naturkatastrophe: eine geo- und hydrochemische Untersuchung der regionalen Arsen-Anreicherung im Grundwasser West-Bengalens (Indien)

Band 30: F. Pujol (2005)
Chemostratigraphy of some key European Frasnian-Famennian boundary sections (Germany, Poland, France)

Band 31: Y. Dikikh (2006)
Adsorption und Mobilisierung wasserlöslicher Kfz-emittierter Platingruppenelemente (Pt, Pd, Rh) an verschiedenen bodentypischen Mineralen

Band 32: F. I. Müller (2007)
Influence of cellulose ethers on the kinetics of early Portland cement hydration

Band 33: H. Haile Tolera (2007)
Suitability of local materials to purify Akaki Sub-Basin water

Band 34: M. Memon (2008)
Role of Fe-oxides for predicting phosphorus sorption in calcareous soils

Band 35: S. Bleeck-Schmidt (2008)
Geochemisch-mineralogische Hochwassersignale in Auensedimenten und deren Relevanz für die Rekonstruktion von Hochwasserereignissen

Band 36: A. Steudel (2009)
Selection strategy and modification of layer silicates for technical applications

Band 37: E. Eiche (2009)
Arsenic mobilization processes in the red river delta, Vietnam: Towards a better understanding of the patchy distribution of dissolved arsenic in alluvial deposits

Band 38: N. Schleicher (2012)
Chemical, Physical and Mineralogical Properties of Atmospheric Particulate Matter in the Megacity Beijing

Band 39: H. Neidhardt (2012)
Arsenic in groundwater of West Bengal

Band 40: X. Tang (2014)
Separating arsenic oxyanions from natural waters for oxygen isotope analysis

Band 41: A. K. Nothstein (2016)
Selenium Transfer between Kaolinite or Goethite Surfaces, Nutrient Solution and *Oryza Sativa*

Band 42: H. Banning (2017)
Selenium Isotope Studies in Plants: Development and Validation of a Novel Geochemical Tool and its Application to Organic Samples